DATE		

Geopolitics of the
Southern Cone
and Antarctica

Geopolitics of the Southern Cone and Antarctica

edited by
Philip Kelly & Jack Child

Lynne Rienner Publishers • Boulder & London

Published in the United States of America in 1988 by
Lynne Rienner Publishers, Inc.
948 North Street, Boulder, Colorado 80302

and in the United Kingdom by
Lynne Rienner Publishers, Inc.
3 Henrietta Street, Covent Garden, London WC2E 8LU

Library of Congress Cataloging-in-Publication Data
Geopolitics of the Southern Cone and Antarctica / edited by
Philip Kelly and Jack Child
    p.   cm.
    Bibliography: p.
    Includes index.
    ISBN 1-55587-105-4 (alk. paper)
    1. Geopolitics—Southern Cone of South America. 2. Southern Cone
of South America—Foreign relations. 3. Geopolitics—Antarctic
regions. 4. Antarctic regions—International status. I. Kelly,
Philip, 1941– . II. Child, Jack.
F2217.G44   1988   88-18080
327.8098'9—dc19   CIP

British Library Cataloguing in Publication Data
A Cataloguing in Publication record for this book
is available from the British Library.

Printed and bound in the United States of America

The paper used in this publication meets the requirements
of the American National Standard for Permanence of Paper
for Printed Library Materials Z39.48-1984.   ∞

Contents

Maps and Tables

MAPS

TABLES

About the
Contributors

César N. Caviedes is professor of geography at the University of Florida. Author of *The Politics of Chile: A Sociogeographical Assessment* and *The Southern Cone*, he is a specialist on geography of western South America and the political geography of the Southern Cone countries.

Therezinha de Castro, professor at the Colégio Pedro II in Rio de Janeiro, is well known throughout the Americas for her geopolitical writings, particularly her research on Antarctica. Among her many publications are *Rumo a Antártica, Atlas-Texto de Geopolítica do Brasil* and *O Brasil No Mundo Atual: Posicionmento e Directrizes*.

Norberto Ceresole is director of the Instituto Latinoamericano de Cooperación Tecnológica y Relaciones Internacionales (ILCTRI) in Buenos Aires. He is author of seventeen books covering political and geopolitical themes.

Jack Child, professor of Spanish and Latin American Studies at the American University, Washington, D.C., was born in Buenos Aires and lived in South America for eighteen years before coming to the United States. His publications include *Unequal Alliance: The Inter-American Military System, 1938-1978* and *Geopolitics and Conflict in South America: Quarrels Among Neighbors*.

Margaret L. Clark is a doctoral candidate in political science at Ohio State University, where she specializes in international relations. Her research interests include international negotiations, international organizations, and peace research.

Martin Ira Glassner, is professor and chair of the Department of Geography at Southern Connecticut State University. His popular text, *Systematic Political Geography*, is now in its fourth edition.

Leslie W. Hepple is professor of geography at the University of Bristol,

England. In addition to his work in geopolitics, political geography, and spatial econometrics, he is author of a historical geography of Northeast England.

Rubén de Hoyos, born in Buenos Aires, is professor of political geography at the University of Wisconsin, Oshkosh. His varied research interests include the Cuenca del Río de la Plata, the influence of church and armed forces on political processes, and the Malvinas/Falklands Islands.

Philip Kelly, is professor of political science at Emporia State University. Coadministrator of the Fitzgibbon-Johnson Image-Index of Latin American democracy, his main research interests include geopolitical theory and the geopolitics of Latin America.

Carlos de Meira Mattos, division general (retired) in the Brazilian army, and professor at Mackenzie University in São Paulo, is one of the most prolific and influential writers on South American geopolitics of the twentieth century. Among his many publications are *Estratégias Militares Dominantes*; *Geopolítica e Trópicos*; and *Uma Geopolítica Pan-Amazônica*.

José Osvaldo de Meira Penna has served most of his life as a career diplomat. At present he is professor of diplomatic history at the University of Brasília and writes frequently for national newspapers. His most recent books are *A Ideologia do Século XX* and *O Brasil Na Idade Da Razão*.

Howard T. Pittman is perhaps the best known U.S. scholar of Southern Cone geopolitics. His numerous publications on the topic reflect his extensive travels throughout the region and his interviews with its leading authorities.

Bernardo Quagliotti de Bellis presently is professor of geopolitics in the Naval War School of Uruguay and editor of the journal *GEOSUR*. A founder and now secretary general of the influential Asociación de Estudios Geopolíticos y Relaciones Internacionales and the Instituto Uruguayo de Estudios Geopolíticos, he has served also as a journalist, as secretary for the majority in the national government, and as adviser in international relations to Uruguay's senate.

Roberto Russell is coordinator of the international policy area of the Facultad Latinoamericana de Ciencias Sociales (FLACSO) in Buenos Aires, and professor of international relations theory at the Foreign Service Institute of Argentina. His areas of research and publication include international relations, U.S.-Argentine relations, and U.S. policy toward Latin America.

Julia Velilla de Arréllaga is director of the Instituto Paraguayo de Estudios Geopolíticos e Internacionales of Asunción. A noted historian of Southern Cone geopolitics, she includes the influential *Paraguay: Un Destino Geopolítico* among her publications.

South America

Geopolitics, Integration, and Conflict in the Southern Cone and Antarctica

Philip Kelly

Jack Child

This book comes at a time when North American interest in geopolitics has revived. Owing much to Henry Kissinger's frequent use of the term, geopolitics is no longer associated with the distortions of pre–World War II realpolitik. Instead, it has become a positive expression linked to foreign and security policies and to intercontinental strategic relationships.

In South America, geopolitics has remained a popular concept throughout this century (Hepple 1986, 522):

> It should be emphasized that the sequence of decline and revival discussed here applies to geopolitics in North America and Europe, but not to the extensive South American geopolitical literature. The latter tradition has flourished and expanded throughout the period (Child, 1985), with considerable political impact . . . [which] has been largely unknown outside South America and has had little impact on geopolitical thought outside that region.

During the past decade, nevertheless, a connection between northern and southern geopolitical thinkers began gradually to be forged and has steadily expanded to the present day.

The authors perceived the need for a book like this for several reasons. First, we in North America respect the scholarship on geopolitics of our Southern Cone colleagues, for they have stimulated and strengthened our own research. Unfortunately, our contacts with them and their work have been more limited than we would desire. Second, we believe our thinking on Southern Cone and Antarctic geopolitics has matured to a point where it would be useful to integrate our perceptions more fully with those of South American and British academics also concerned with the topic. In both regards, it seemed to us that an edited book such as this would enable a geopolitical cross-fertilization to continue among researchers of three continents.

1

Third, we wanted to share what we have learned with our readers, to whom we offer geopolitical descriptions and insights of a world region that is increasingly independent from Great Power politics, and of peoples who are wary of potential territorial disputes, but who likewise are intrigued by possible benefits of intraregional integration. In sum, the Southern Cone is an area with a rich international heritage, a present in transition, and a future in which a greater impact in global affairs could well emerge.

Latin America in its geopolitical complexity can be visualized within a subregional context, the two main areas being Middle America (Mexico, Panama, Central America, northern South America, the Caribbean islands and rimlands) and the Southern Cone of South America (Brazil and Peru, south to Antarctica). Throughout its Columbian history, Middle America has regularly alternated in international status between a shatterbelt configuration (that is, a zone of Great Power contest) and a Great Power's sphere of influence. The immediate danger from Middle America is potential for disputes occurring in its domain to escalate into superpower confrontation, a characteristic of its present shatterbelt condition. In global affairs Middle America will react, not act; it will continue to be divisive and depressed and unstable; and its dependency will remain fixed to the charge of some outside Great Power (Kelly 1986).

In contrast, the wealthier, less fragmented Southern Cone is today a more independent region in global geopolitics, quite isolated from Great Power strategic conflict and interference, beginning a promising economic integration, and potentially able to exert a much stronger international influence. We see this autonomy and growth—and a new era of regional political harmony—as likely to persist, particularly in light of the current rapprochement between Brazil and Argentina and the emergence of Brazil as continental integrator and peacemaker.

As defined in this account, the Southern Cone encompasses South American territorial and maritime spaces below 10 degrees south latitude, that is, roughly south of the Amazonian tropics. Five countries—Argentina, Chile, Uruguay, Paraguay, and Bolivia—lie entirely within its bounds, and a sixth area—the vastness of southern Brazil—adds a further important sector to the region. Adjacent ocean spaces of the South Pacific, the South Atlantic, and the Antarctic are likewise a part of the Southern Cone designation. We chose to exclude Peru because its location is peripheral to our interest and its geopolitical involvement for the most part has been more internal than external in scope.

We define geopolitics as the impact on foreign and security policies of certain geographic features, the more important among these being locations among countries, distances between areas, and terrain, climate, and resources within states. Geopolitics might also be described as the relationship between power politics and geography. The usefulness of geopolitical analysis derives in part from the formulation of broad linkages or theories among these geographic features and policies, linkages that bring insight to international relationships. Geopolitics, we believe, represents one method for studying foreign and strategic affairs, and it relates as much to planning for peace as it does to military involvement.

At the same time we cannot ignore other perspectives and harsh critiques of Southern Cone geopolitical thinking. A particularly unsettling aspect has been the linking of geopolitics to national security programs and doctrines of military/dictatorial states and to a series of repressive regimes in South America over the past quarter-century. However, the authors feel that this negative current of geopolitical thinking has become increasingly anachronistic as the Southern Cone consolidates its democratic traditions and emphasizes the geopolitics of cooperation and integration.

Each chapter author has been carefully selected. Of the sixteen, seven (Ceresole, de Castro, Meira Mattos, Meira Penna, Quagliotti de Bellis, Russell, and Velilla de Arréllaga) reside in the Southern Cone. Two U.S. citizens, Caviedes and de Hoyos, have origins in and long-standing associations with Chile and Argentina, respectively. Hepple provides a British view of the 1982 Malvinas/Falklands dispute in the South Atlantic. The remaining five authors (Child, Clark, Glassner, Kelly, and Pittman) are North Americans. Each of these authorities is noted for his or her research on a particular dimension of the topic, and it was our task as editors to locate common denominators that would combine their expertise into a comprehensive geopolitical vision of the Southern Cone and Antarctica.

Our initial instructions to the contributors were that each was to analyze some or all of three topics: (1) the prospects for conflict among states in the Southern Cone; (2) the potential for regional economic and political integration; and (3) the role of Brazil toward influencing levels of conflict and/or integration in the Southern Cone. Being independent-minded professionals, the authors took differing paths. Some maintained previous stances and subregional specializations; others explored new ground and interpretations. Consequently, the chapters variously speculate, evaluate policy, focus on strategic considerations, or speak about geopolitical conceptualizations. Nevertheless, in the aggregate they cover all relevant areas, both spatial and intellectual.

In this connection it is instructive to assess the relationship between geopolitical thinking and interstate conflict in the region. A decade ago there seemed to be strong linkages between such conflicts and geopolitical thinking, especially in relation to the national security policies of military states of the Southern Cone. In fact, it was possible to identify a balance of power system in the Southern Cone which was based on a series of geopolitical "laws." Were this balance of power system to break down, one could envision a number of conflicts (or increased tensions) breaking out.

This geopolitical framework took note of the understandings or loose alliances between two pairs of key Southern Cone nations which had no contiguous borders: Chile-Brazil and Argentina-Peru. It also was related to a series of bilateral strains involving those nations with borders: the historic Argentine-Brazilian rivalry; Argentine-Chilean strains (Beagle Channel islands, Patagonia, Antarctica); Chilean-Peruvian tensions over the War of the Pacific; Bolivian-Chilean difficulties stemming from Bolivia's search for a sea outlet; and Peruvian-Ecuadorean confrontations over border-delimitation problems stemming from the 1941 war. In the 1980s, this old geopolitical balance-of-power argument has lost force as many of these rivalries have softened and given way to new forms of cooperative geopolitics. Thus, the resolution of the Beagle Channel problem in 1985 (with Vatican mediation), the signing of numerous Argentine-Brazilian agreements, and the peaceful passing of the centennial of the War of the Pacific have all contributed to the new emphasis on cooperation, integration, and development.

However, not all of the old geopolitically rooted conflicts have been eliminated, and there is still the possibility of a renaissance of tensions over issues of sovereignty and resources. This is especially evident in Antarctica and the southern waters that separate the Antarctic Peninsula from South America. As brought out in several of the chapters that follow, the multiple pressures of ambiguous Antarctic sovereignty, competing claims, suspicion of resource exploitation, and the psychological deadline of 1991 (when the Antarctic Treaty may be reviewed and possibly revised) all combine to suggest that old continental geopolitical rivalries and conflicts might find an outlet in Antarctica. At the same time, the very authors who note this pessimistic scenario generally acknowledge that Antarctica also provides fertile ground for the further growth of South American cooperative geopolitics. The outcome is obviously still in doubt and will probably remain so for some years.

On the question of whether the Southern Cone would experience violence in the near future, a rough consensus emerges among the chapter writers. With the exception of Pittman and de Hoyos, most authors believe that the region is moving toward a harmony of national interests. This might reflect the rather widespread shift to democratic forms of government. It could also result from the debt-repayment problem or from stagnating national economies. Likewise, the Malvinas/Falklands conflict dampened enthusiasm for aggressive intraregional policies, since most republics (the exception being Chile) felt themselves abandoned by the United States and exposed to NATO and even to Soviet intervention in the South Atlantic. Consequently, chapter authors from both North and South America urge continuation of the emerging Argentine-Brazilian rapprochement and a common political front in Antarctica and the adjacent seas toward the northern world, and they are optimistic about the abilities of Southern Cone members to resolve territorial disputes within the present situation and to preserve a consistent position against outside forces.

A critique of the geopolitics of integration or cooperation is also important to this book because it is a basic element in the apparent shift from the hostile geopolitics so characteristic of the 1960s and 1970s to the more harmonious geopolitics of the 1980s. In the earlier period Argentine writers in particular alleged that there were deep divisions in the Southern Cone, which, they argued, stemmed from the role of Brazil in the region. For many of these analysts Brazil's close politico-military and economic alliance with the United States meant that Brazil was not truly a Latin American nation and that Argentina had to resist Brazil's emergence as a regional and world power in order to block U.S. domination of the area through the Brazilian surrogate. These same geopolitical writers later noted the strains between Brazil and the United States in the Carter years and felt that as Brazil broke past close ties with its northern partner it was ready to be welcomed back into the Latin American fold. This, in turn, cleared the way for the Argentine-Brazilian rapprochement, analyzed by several of this book's authors, that seems to be the harbinger of a new era in Southern Cone geopolitics. The proponents of this particular view regard integration with an enthusiasm that sometimes borders on euphoria.

This suggests that an objective critique of integration geopolitics might lead to the conclusion that its potential and past successes have been exaggerated. Latin American integration, after all, is nothing new. It can be traced back as far as the 1826 Amphyctionic Congress convoked by Simón Bolívar, and to a series of Pan-Hispanic American conferences in the middle and late nineteenth century. But these early attempts at forging Latin American solidarity generally failed, as nationalism and parochial interests overwhelmed the drive for integration.

The common factor in successful attempts at regional unity has frequently been the United States, both in a positive sense as the agglutinator, and in a negative sense as the adversary or threat against which Latin America must organize. As an example of the positive role of the United States we might cite the Pan American thrust, which brought together the republics of the hemisphere in a late nineteenth-century entity that later evolved into the Organization of American States. The negative role of the United States in hemisphere integration can be seen in the Latin American reaction to U.S. policies in the 1982 Anglo-Argentine War, when a prominent South American noted that U.S. support for Margaret Thatcher had achieved what no Latin American had ever accomplished: the unity of Latin America. Thus, any critique of Latin American integration must include both the hopeful optimism so prevalent among its proponents, and the more tempered realism informed by past failures at unity.

In like manner, most of our authors note progress in integration among Southern Cone nations. Indeed, this theme dominates a good many of the chapters and is perhaps the geopolitical factor most stressed in the book. The La Plata watershed, particularly the river's mouth, and the hydroelectric power sources (especially Itaipú) appear as focal points for plans for coordinated development. Surprisingly, we find little skepticism (the main exception being Glassner's chapter) about the likely success of integration, even in the face of the region's difficult terrain, scarcity of resources, and tumultuous political past. Perhaps the authors remain supportive of the integrative ideal because they believe that such ventures must at least be attempted and that their accomplishment could render significant regional benefits. Placed within a broader geopolitical framework, without integration of some substantial extent, the Southern Cone could remain fixed to the world political and economic periphery, manipulated by external influences and divided internally by territorial and power disputes.

Another important topic of this book is the role of Brazil in the Southern Cone. Brazilian power and central continental position could soon represent such a continental impact that it could strongly influence, if not dictate, outcomes to most conflicts as well as to most integrationist attempts in South America. Brazil leads other continental states in land and population size and in most areas of industry, technology, transportation, natural resources, government efficiency, and military power projection. It also holds mastery over the Amazon basin and shares control over the La Plata watershed.

Several stances in particular characterize Brazil's relations toward the Southern Cone (Kelly 1987). For example, much of the earlier territorial expansion of Brazil transpired according to an *imperialist* foreign policy, marked by a forceful seizure of land from adjacent states. Peripheral Spanish nations continue to fear a resumption of Brazil's former aggressiveness. Also in earlier decades, Brazil actively sought a Southern Cone *balance of power* that placed particular emphasis upon maintaining Chilean independence against threats from Peru and Argentina. This balance-of-power policy sought to prevent Spanish encirclement, to assure the independence of the buffer states (Bolivia, Uruguay, and Paraguay), and to allow pursuit of major power status in world politics. Nevertheless, the balance created the potential for serious conflict escalation, in particular with Argentina, either through rivalry in the buffer areas or in conventional or nuclear arms races.

Brazil frequently has been accused of *surrogate* ties with the United States, a strategic collaboration allegedly bringing Brazil international status, economic assistance, and protection from her northern benefactor. Usually, this viewpoint portrays Brazil intervening against subversion and conflict in the Southern Cone but remaining aloof from regional integrationist ventures. Another common characteristic of Brazil's foreign policy is its striving for *great power status*, a special "destiny" derived from its unique Portuguese heritage, natural wealth, large continental space, and position between the Southern Cone and northern continents.

More recently, Brazil appears to have taken a regional *hegemonistic* policy as stabilizer and integrator in Southern Cone affairs. As the strongest military and economic power of the area, Brazil is not only the primary designer and enforcer of continental "rules of the game," but also the guarantor of territorial integrity and opposer of serious conflict within or among the region's states. Finally, economic difficulties seem to have convinced Brazilian policymakers that Southern Cone *integration* offers a promising path to national recovery and growth. For this, the country's leaders apparently believe Brazil must integrate its hinterlands with continental interior and coastal areas and expand trade with Third World countries, including those of their own hemisphere. Such a role stresses cooperation with South American neighbors on various projects in the Amazon and involving La Plata hydroelectric power, all designed to bring benefits that transcend national borders.

The growing power of Brazil and the country's potential for territorial expansion against Spanish America elicited minimal apprehension from our chapter contributors. Likewise, aggressiveness is largely absent from the chapters by the Brazilians. Brazil is portrayed here as a respected and full member of the Southern Cone, and, likewise, as a positive contributor to integration, a protector of the South against northern intrusions, a frontier stabilizer, and a low-key, cautious player of regional diplomacy in general. Will this image persist, or will suspicion reemerge among the Spanish speakers and aloofness among the Portuguese speakers? There seems little— at least as seen by the authors of this book—to indicate that Brazil will not continue its present path of regionalism.

Another important geopolitical aspect covered in this book is the Southern Cone's strategic projection, that is, the region's reaction to or impact upon other continental and maritime spaces. The primary thrust of strategic geopolitics for the past two hundred years has focused on a balance of power in Eurasia, that great expanse of territory from Europe through Asia which encompasses nearly two-thirds of world population and on-land natural resources. A power equilibrium in Eurasia allegedly secures peace in Asia and Europe as well as in peripheral zones in Africa, the Americas, and Australasia. The security policies of the United States, for example, aim to project military power onto certain Eurasian areas when necessary so as to assure an equilibrium or favorable balance of power (Posen and Van Evera 1987; Gaddes 1983; Spykman 1942). Hence, this power balance prevents major Eurasian nations from attempting aggression in America.

The Southern Cone essentially is removed from the geopolitics of this preoccupation with a Eurasian balance because of its southern remoteness, its internal economic and political fragmentation, its lack of strategic location and natural resources, and its isolation behind the protected southern flank of the United States. The region's only strategic impact comes from its maritime straits, primarily the Atlantic Narrows, through which flows Middle East oil destined for the North Atlantic, and the Southern passages (the Drake Passage, the Beagle Channel, and the Magellan Strait). Potential strategic importance could someday derive from the Southern Cone's position astride sea lanes between Africa, Antarctica, the South Atlantic, and the South Pacific, if resources from these areas become vitally significant to industrial nations. We are able to see some forecast of this in the contemporary rivalry for possession of Antarctic wealth.

Two final geostrategic comments need also to be expressed. First, the global impact of the Southern Cone could expand if the region becomes integrated, particularly if Argentina and Brazil effectively coordinate their foreign and security policies. A Cold War stalemate in the North could likewise benefit the autonomous southern position. Second, located as it is on the Eurasian periphery, the strategic position of the Southern Cone parallels that of North America in that both areas stake their security on a Eurasian balance. Whether this spacial similarity suggests strategic alliance with the United States, or instead, an independent policy of neutrality must of course be determined by South American leaders.

Concern with broader, more strategic geopolitical features appears in many of the chapters. There is agreement, for instance, that U.S. interest and influence in Southern Cone affairs has clearly declined. Whether or not this has been good for the region, however, brings some amount of sharp disagreement (see the Ceresole and de Castro chapters). A related element is the awareness that the Southern Cone could be on the verge of becoming a more important global strategic force, an area increasingly able to exert impact upon northern balance of power struggles. Many of our contributors feel that this strategic importance is dependent in large part on a continuation of both unity in the Southern Cone and cleavages in the northern hemisphere. Some authors also draw attention to the strategic value of southern islands and straits (the Malvinas/Falkland Islands, the Drake and Beagle passages, the Atlantic Narrows, and Antarctica itself, for example), particularly in cases of global warfare, of blockage of the Suez or Panama canals, and of the persistence of Middle East oil shipments through the South Atlantic to North America and Western Europe. Contesting this latter assertion, Child and Hepple hold that the importance of these choke points is exaggerated and that northern power rivalries are likely to remain located above the equator.

The few disagreements among the authors concern North/South political orientations. For example, Child and Hepple clearly depart from de Hoyos, de Castro, and Meira Mattos in assessing the strategic importance given the South Atlantic and the Antarctic by northern powers. The former maintain the relative strategic insignificance of the South Atlantic; the latter assert a heightened strategic worth of southern oceans occurring in future decades. One could also contrast Pittman's and Russell's depiction of democracy in the Southern Cone. In addition, although Kelly and Ceresole agree on the definition and importance of the shatterbelt phenomenon relative to the Southern Cone, in practical application Kelly sees an absence of a southern shatterbelt, whereas Ceresole declares its presence.

A major problem with the academic study of geopolitics, and one this book confronts, is that geopolitics is conceptually and theoretically vague. An adequate geopolitical model does not exist. Such a model would possess clearly defined concepts and rigorously tested theories, show linkages among such concepts and theories for analyzing foreign policies and international events, and enable its users to interpret and predict a wide scope of international phenomena based upon a consistent, well-integrated assortment of geopolitical conceptual linkages. A model is necessary to an application of geopolitics to foreign and security policies and, in our case, the Southern Cone, because it helps label what is or is not geopolitical, it assists in locating relationships among events, concepts, and theories within a common framework, and it merges all of these aspects together into a much larger and cohesive general approach to the study and the practice of international politics.

What do the authors accomplish in this book by way of advancing the geopolitical approach to international affairs? We believe that they consistently and at times inventively apply recognized geopolitical concepts and theories to Southern Cone events (Mackinder, Mahan, Golbery, Beaufre, and Cirigliano, among others, are cited). Perhaps their greatest contribution is their insight into strategic linkages between the position and resources of the Southern Cone and other global spaces that impact on the region. We applaud their dedication to "middle-range" levels of theory, this being a necessary step the study of geopolitics must now take.

Finally, what geopolitical alternatives appear most likely for the Southern Cone and Antarctica in the near future? We are able to visualize three. One possibility is substantial integration, which is dependent on receptive regional and global economic and political climates and on economic pay-offs for both Brazil and Argentina. If such integration succeeds, regional peace, economic growth, strategic autonomy, and heightened international stature will transpire. A second is that increased levels of conflict could occur, with a revival of territorial disputes; a return of Brazilian aloofness, surrogate image, and rivalry with Argentina; a breakdown of integration and autonomy; and a possibility of a Southern Cone shatterbelt. A third scenario is eclectic, a mixture of the first two. Scattered attempts toward integration, occasional conflict over disputed territories, and uneven national growth among member states may take place, features resembling Southern Cone history during the first six decades of the present century.

We have organized this book into six parts. In two initial chapters, Caviedes and Pittman provide a geopolitical context for the entire Southern Cone. In Part Two on Argentina, Ceresole and Russell describe a nationalist and a policy orientation respectively. Part Three covers Brazilian geopolitics from three varying perspectives: the historic (Meira Penna), the inter-American security (de Castro), and the traditional (Kelly). In Part Four the buffer states of Uruguay, Paraguay, and Bolivia are examined, with a heavily integrationist orientation, by Quagliotti de Bellis, Velilla de Arréllaga, and Glassner. A focus on Chile (Pittman) in Part Five is followed in Part Six by chapters devoted to the Antarctic (Child and Clark), the strategic importance of the South Atlantic (Meira Mattos), and the Malvinas/Falklands controversy (Hepple and de Hoyos).

part one

BACKGROUND

The Emergence and Development of Geopolitical Doctrines in the Southern Cone Countries

César N. Caviedes

Doctrines that emphasize the politicizing of a territory or insist on the strategic value of a space have developed and found their most elaborate expressions in the countries of the Southern Cone. Special historical circumstances contributed to the emergence and continuation of these doctrines, such as territorial ambiguities generated by the imprecise colonial claims that subsequently prompted numerous boundary conflicts among Latin American nations, or the perception of geographical space as the ground where youthful and aggressive nations tested their strength.

Obviously, the mutual apprehensions and tensions that expansionism and boundary disputes gave rise to favored the appearance of an acute geopolitical awareness among the Southern Cone nations, giving an exaggerated role to such geopolitical constructs in the relations of the Southern Cone states among themselves and outside the continent. Fears of surreptitious penetrations by aliens of boundaries in remote areas have elicited geopolitical doctrines that emphasize territorial integrity, foster national security programs, and oppose the development goals of bordering states. Such ideologies are defended with equal vehemence by militaries and well-educated civilians and a considerable intellectual effort has been applied to their elaboration, reinforcing the degree of tradition and credibility that geopolitical thought still enjoys in the Southern Cone countries.

THE INHERITANCE OF IMPRECISE COLONIAL DEMARCATIONS

The bases of most friction between Latin American nations lay in the vague boundary lines between the numerous *gobernaciones, audiencias,* captaincies, and viceroyalties through which the Spanish colonies were administered. At the root of the territorial disputes was the antagonism that separated Spain and Portugal, the contending imperial powers in South America. After the Treaty of Tordesillas (1494), which determined the extent of the divided territories, both nations embarked on campaigns of discovery and occupation in unexplored territories that were neither properly surveyed, nor even rudimentarily mapped. Vast as these South American lands were, the possibilities of conflict between the two contending powers were rare, but in certain areas, such as the territories of Misiones and the estuary of the Río de la Plata, armed encounters between Spaniards and Portuguese were frequent throughout colonial times.

Interpreting to their advantage the demarcations prescribed by the Treaty of Tordesillas, the Portuguese claimed the northern shore of the La Plata estuary and justified the establishment of a military outpost at Colonia de Sacramento, just across from the Spanish settlement at Santa María de los Buenos Aires. The enclave served as bridgehead into the interior of the La Plata Basin and as a point from which to monitor the Spanish presence in Buenos Aires. These intentions reveal the Portuguese awareness of the estuary's significance as the entrance into the vast La Plata lowlands, a region where the riches of the "White King of the Silver Mountains" were supposed to lie (Samhaber 1946, 167–188; Hanson, 1967, 197). The intrusion of the Portuguese into the estuary and their destabilizing efforts on the Platine provinces continued into republican times and led to the secession of Uruguay, the "Eastern Seaboard," from Buenos Aires' rule. The Argentinian resentment against imperial Brazil changed little after independence from Spain, and today an overwhelming number of Argentines still harbor an entrenched suspicion of Brazil's aggressive expansionism. Another example of perceptions about the key role to be played by the La Plata basin in a future confederation of South American Spanish-speaking countries explains the drive for independence by Paraguay—another country split from the former Viceroyalty of Río de la Plata—in search of a prominent role in the center of the platine lowlands. No less decisive for the further dismemberment of the former Spanish empire were the ambiguities about the control of the territory of Misiones, culturally dominated by Paraguay, contested by the Portuguese after the removal of the Jesuits, and placed under the jurisdiction of the Viceroyalty of Buenos Aires in 1776. Under the sovereignty of Argentina, the "salient of Misiones" (as it is called in the works of several Argentine geopoliticians) continues even today to be a sensitive issue in the relations among Argentina, Brazil, and Paraguay.

The persistence of territorial claims among Southern Cone countries that emerged in colonial times is further illustrated by the projection sought by Argentina and Chile into the Cape Horn area and the Antarctic. The origin of this drive lies in the early assumption that a "passage" connected the Atlantic

Ocean with the "Sea of the Indies" at the southern extreme of South America. When Magellan discovered the passage in 1521, there was little doubt in the minds of Spanish cosmographers that the southern shore of the strait was the northern edge of a continent stretching to the austral pole. Based on this assumption, Pedro Sánchez de Hoz obtained in 1539 a royal letter that authorized him to take possession of *terra australis*, the name given to the territories located south of the Strait of Magellan (Tanzi 1985).

Not even the discovery of Cape Horn in 1622, which revealed the insular character of Tierra del Fuego, convinced the Spanish cosmographers that there was no *terra australis*, and all through the colonial period the concept persisted that the Magellan Strait, as well as the Sea of Drake, were only passages connecting the Atlantic Ocean with the Pacific Ocean, "the Spanish Lake" (Spate 1979). Consequently, the imperial authorities recommended occupancy of these remote "passages" by Spanish garrisons and colonists in order to foil the growing attempts of English, French, and Dutch privateers to intrude into the Pacific and disrupt the colonial trade routes there. This policy implies the Spaniards' recognition of the geostrategic significance of the southern interoceanic passages, an awareness that was to become even more compulsive when Argentina and Chile emerged as independent nations. Thus, the contemporary efforts of both countries to gain uncontested control of the Strait of Magellan, of the Beagle Channel, and of Cape Horn, and their ultimate desire to extend their dominance over the South American quadrant of the Antarctic for geopolitical reasons, must be traced back to their colonial origins in order to be properly understood. The imprecise territorial boundaries in the southern part of the Viceroyalty of the Río de la Plata and the kingdom of Chile also caused disputes between Argentina and Chile for the control of Patagonia, the Strait of Magellan, and Tierra del Fuego during the nineteenth century. Similar vagueness concerning the colonial limits of the Viceroyalties of Peru and the Río de la Plata and of the kingdom of Chile prompted territorial and boundary disputes that have strained relations between Bolivia and Chile, Argentina and Chile, and Bolivia and Paraguay since the last century.

These antecedents demonstrate that the spatial considerations and the strategic postulates contained in the geopolitical constructs of Southern Cone countries and their neighbors are the perpetuation of geographical perceptions raised in colonial times or the inheritance of vaguely defined territorial boundaries. Certainly, these explanations are not exclusive to the nations of the Southern Cone; they apply also to the boundary conflicts among other nations of the continent. The only difference is that in the latter these motives were not the foundations of obsessive geopolitical programs.

THE COMMON CHARACTERISTICS OF
COLONIZATION AND TERRITORIAL EXPANSION IN
THE COUNTRIES OF THE SOUTHERN CONE

Dating back to the formative years of the new republics is the conviction held by South American leaders that national grandeur can be achieved by means of a well-conducted occupation of unpopulated territories, by the incorporation of contiguous disputed lands, and by the development of a military force capable of realizing the two previous objectives. Thus, as soon as the countries effected territorial occupation, international circumstances provided the means for them to undertake expansionist ventures beyond the limits inherited from the colonial period.

Once the flaring anarchy of the postindependence years was overcome, Argentina, Chile, and Uruguay encouraged the planned immigration of Europeans into empty national spaces. In this manner the Southern Cone countries tried to emulate the populating strategies of the United States of America, which they perceived as the pathway to progress. Governmental policies of territorial expansion were carefully executed with the purpose of claiming vacant lands or occupying resource frontiers of potential value. Another motivation for settling empty spaces was the desire to establish colonies and military posts in places that were perceived as strategically significant.

Having reached internal cohesion and institutional stability before Argentina, Chile could move ahead, in 1843, in its attempts to claim and actively occupy territories in Patagonia, Tierra del Fuego, and the Magellan Strait. With this advantage, Chile could also claim sovereignty over trans-Andean Patagonia, thus starting a dispute the consequences of which persist until today. In addition, the entrepreneurial spirit of Chilean agriculturalists and merchants pushed them to sell wheat, tallow, and wine as far away as California, Australia, and Polynesia, thus promoting the emergence of a national merchant marine that took not only the Chilean flag to Polynesia, but also Chilean gold pesos (Bunster 1964, 37–53). Chile's startling commercial successes in the Pacific towards the middle of the nineteenth century gave rise to the Chilean perception that, in the South Pacific, it was as great a power as Great Britain or France and led to the ideology of Chile's manifest destiny in the South Pacific. This feeling, added to the control of the routes of access from the Atlantic to the Pacific, enhanced Chile's conviction of its maritime superiority. Moreover, the relentless flow of Chilean miners, supported by national and foreign capitalists, into Bolivian territory north of parallel 24 degrees south latitude, was viewed as an expression of entrepreneurial dynamism and used as the justification for the war against Bolivia and Peru in 1879. At the end of the war Chile annexed the Bolivian territories on the Pacific coast and the southern provinces of Peru, thus mustering an impressive frontage onto the Pacific Ocean that enhanced the image of the country as a maritime power.

In Argentina the dissensions between Buenos Aires and the restive provinces that had once been integral parts of the La Plata viceroyalty resulted not only in the secession of Uruguay, the independence of Paraguay, the loss of the Malvinas, and the anticipation of Chile in the occupation of the Strait of Magellan, but it also plunged the country into successive periods of anarchy and dictatorship that persisted up to the fall of the dictator Juan Manuel de Rosas in 1852. It was after the War of the Triple Alliance against Paraguay (1865–1869) that the tide reversed for Argentina: the territory of Misiones returned to the sovereignty of Buenos Aires and a large segment of the Southern Chaco was annexed. Free of the tensions in the upper La Plata Basin, Argentina had fewer military commitments to concentrate on and directed its energies to wresting the Pampa away from the Indians through the "Campaigns of the Desert," to reinstating its claims to eastern Patagonia by forcing Chile into accepting negotiations over territories that were allegedly Chilean, and to beginning its long, if unsuccessful, campaign for the recovery of the Malvinas/Falkland Islands (Hoffman and Hoffman, 1982).

Brazil, the other winner in the war against Paraguay, expanded westward at the expense of Paraguay and, later, of Bolivia, through the annexation of the Acre territory. After this expansion Brazil did not enlarge its territories but, under the apprehensive eyes of its underpopulated neighbors, conducted an aggressive mobilization of its living frontiers towards spaces susceptible of occupation.

Lodged between two giant neighbors, Uruguay was to engage in a policy of containment and diplomatic accommodation in order to preserve its territorial integrity and political independence, although its role in the La Plata region appears often subordinated to the dictates of Argentina.

In view of the consolidated prestige and international weight of Argentina and Brazil in the southwestern Atlantic and La Plata Basin, and of Chile in the southeastern Pacific, it is understandable that territorial expansions by means of military campaigns and diplomatic pressures justified the existence of powerful military establishments and legitimized the use of force in international conflicts. The end of the century witnessed a growing arms race in Argentina, Brazil, and Chile that brought Argentina and Chile to the brink of war in 1899, and has since fueled the mutual distrust between Argentina and Chile and between Argentina and Brazil, regarded as an ally of the Chileans.

In this context expansionism was considered an expression of the virility of youthful nations. In fact, very few individuals or institutions in these countries dared to question the morality of territorial gains by means of force and at the expense of less aggressive countries, and not even the European or North American powers voiced any criticism, engaged as they were in similar practices.

Obviously, territorial expansion by means of armed interventions had to engender militarism in the countries involved and led to the development of nationalist ideologies capable of keeping alive the ideas of virility and manifest destiny that the military conquests had spawned.

NATIONAL PROGRAMS
AND CONTEMPORARY GEOPOLITICS

During the nineteenth century the expansion of the Southern Cone states beyond the traditional boundaries inherited from colonial times led to collisions of interest, particularly over territories in peripheral regions of unclear demarcation. Further, to the simple cases of concurrent territorial appetencies another element for international conflict was added, namely, the desire to project spheres of action into outward areas that were visualized as domains for international policies.

In Chile, particularly since the annexation of Easter Island in 1888, the eastern Pacific was envisioned as the oceanic realm into which the country had to project itself. To police that ocean and keep the Argentines confined to the Atlantic, a sizeable and expensive navy was built towards the end of the nineteenth century. These measures enhanced the idea of manifest destiny of Chile in the sea that became the keystone of Chile's geopolitical projects. In addition, Chile sought to expand from its bases on the Strait of Magellan and in Tierra del Fuego into the southern Atlantic and, ultimately, into the South American quadrant of the Antarctic. In his literary work *Tierra de Océano*, the prestigious author Benjamín Subercaseaux explicitly stated the thoughts common among the Chilean intelligentsia about the manifest destiny of their nation. He felt that the heroic pioneer past of Chile in the South Pacific had to be reactivated in view of the postwar emergence of the Pacific as a vital sphere of action in the search for agricultural commodities and sea and mineral resources by the United States, the Soviet Union, and the People's Republic of China. Should the country miss this opportunity by not developing a modern merchant fleet and maintaining the role of primacy in the southeastern Pacific, then Chile was destined to become a third-rank nation (Subercaseaux 1951, 15–17). These prophetic and enlightening words were penned in 1946 by a writer without geopolitical schooling, but capable of foreseeing the future significance of a Pacific realm caught in the power struggle of the United States, the Soviet Union, and China. He hinted at the opportunities that his country had in the vast Pacific Ocean if only it could respond to the challenges that the Great Ocean was offering the promising maritime destiny of Chile. However, due to the progressive economic deterioration of the country during the first half of this century, these issues had become more nostalgia than reality and were again elaborated upon and given a geostrategic connotation by Ramón Cañas Montalva, commander in chief of the army in 1946, and former director of the Military Geographical Institute of Chile. In this capacity he founded, in 1948, the *Revista Geográfica de Chile: Terra Australis,* in which he published numerous articles about the geopolitical programs of Chile. Most of his writings were aimed at unraveling the designs of continental primacy held by Argentine generals and at counteracting the politics of international projection pursued by Juan Domingo Perón during his rule (1945–1956). Cañas Montalva advocated a reestablishment of Chilean preeminence in the South Pacific and insisted on securing control of the Beagle Channel and Cape Horn as bridgeheads for Chilean claims in the Antarctic (Cañas Montalva 1954). No major elaborations on European geopolitical doctrines are evident in his articles and the ideas about the maritime destiny of the country lack originality; still, given his high status in the army and his role as speaker of nationalist postulates, the relevance of Cañas Montalva lies in his having managed to keep alive the concern for Chile's role in the South Pacific and in the southern tip of the continent at a time when Argentina was aggressively pushing its own presence in these regions.

Further intellectual development and a sense of reality concerning Chile's safeguarding the use of the Strait of Magellan, the Beagle Channel, and the Sea of Drake by the western democracies in case of global war or conflict with Argentina was demonstrated in the *Memorial del Ejército* by such military writers as Ramón Salinas F. (1947), Julio Campos S. (1950), and Jorge Vallejo (1950), as well as in the pages of *Revista de Marina* by such naval officers as Hernán Cubillos L. (1950), and Arturo Troncoso D. (1961).

Similar insistences on Chile's protective role concerning the passages between the Atlantic and the Pacific, the projection of the country into the southeastern Pacific, and the route to the Antarctic continued to emerge in the 1950s and 1960s; however, they were but variations on an old theme. Nevertheless, in the late 1970s many of these postulates previously advanced by Chilean geopolitical writers became urgent national issues when the compelling power of Argentine geopolitical demands provoked the crisis over control of the Beagle Channel that brought Chile and Argentina to the brink of war (Guglialmelli 1978). Once that emergency passed, the expansion by the state airline of the Pacific route to Easter Island and Tahiti—and even Fiji and Australia in the early 1980s—revived past dreams of Chile's becoming one of the leading trading nations in the South Pacific.

These all-but-extinguished dreams of grandeur still surface in incidental writings of professional geographers. Ricardo Riesco, a graduate of the University of Bonn, exemplifies the views that circulate among self-professed geopoliticians. Although Riesco does not advance anything modern in geopolitical theory (he makes continuous references to Ratzel and Mackinder), some of his ideas about the valorization of particular geographical spaces, in the context of a new geopolitics of the nuclear and space age, introduce elements of novelty into the outdated geopolitical constructs held by Chilean generals and diplomats. According to Riesco, the Pacific Ocean continues to be the center of political action in the contemporary world, and it is only appropriate for Chile to be one of the managers of the circulation and use of the resources of the South Pacific. (See Map 1.1). Since the nations on the western margin of that ocean—especially China, Japan, the Philippines, and Korea—are overpopulated and either communist or on the verge of falling under communist domination, the underpopulated Latin American nations of the eastern rim should be on the alert and prepare themselves for a hypothetical onslaught of Asian nations seeking *Lebensraum* in South America (Riesco 1980, 54) Furthermore, using the idea of the virtual "narrowing" of the open oceans created by the claims of territorial seas following the last Conference on the Law of the Sea, Riesco visualizes "two straits" connecting the North with the South Pacific: one located between the Tuamotu Archipelago and Easter Island, the other between Easter Island and the Chilean coast. Chile should police these "straits" since they allow foreign nations from the North Pacific to gain access to the Southeastern Pacific and, ultimately, to reach the South American quadrant of the Antarctic, an area which is partially considered the patrimony of Chile. In this view, the role assigned to Easter Island is also crucial. Citing the circumstance that a few years ago NASA expressed an interest in extending the runway of the Matavery airport so that it might possibly be used for emergency landings of U.S. space shuttles, Riesco calls attention to the international significance of that island and raises the question of its precarious attachment to Chile (Riesco 1980, 52). Having underlined the significance of space travel in the future, he foresees the use of the Antarctic as a point of departure for future space-age expeditions due to the clearness of its atmosphere and the absence of radiomagnetic interferences which are so abundant in the populated areas of the Northern Hemisphere (Riesco 1984, A2). Concerning South American relations, he insists that the conflicts that smolder between Chile and Argentina because of their rivalry for the southern tip of the continent and the Antarctic should now be abandoned and replaced by cooperation between the two in an effort to present a firm stand against the intrusions of foreign powers and noncontiguous nations—Brazil included—in the exploration and survey of the resources of the Antarctic. In the opinion of Riesco the future international relations in South America will not be ruled by chance alliances of countries based on revenge or unsatisfied irredentism, but by the confrontation between "Tropical America dark and indigenous, and Extratropical America, white and europeanized [*sic*]." This will be the consequence of Brazil's international politics of rallying countries like Peru, Bolivia, and Ecuador around the development of an Amazonian pole of convergence, whereas Argentina and Chile are increasingly agreeing in their desire to keep other nations out of the southern oceans and in their search for common grounds in the defense of their exclusive rights to the Antarctic (Riesco 1986, A2).

Map 1.1   The Role of Chile in the Newly Developing Geopolitics of the
Pacific Ocean

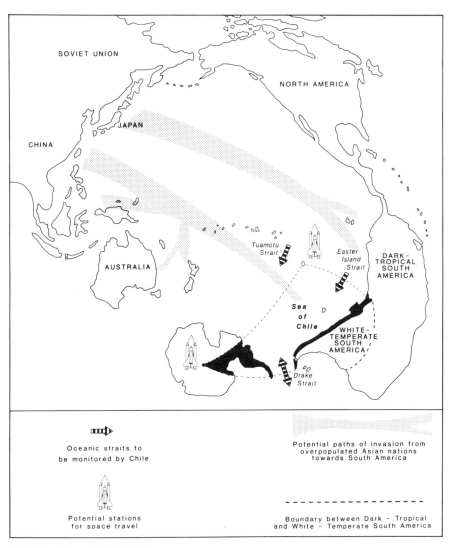

SOVIET UNION

NORTH AMERICA

JAPAN

CHINA

AUSTRALIA

Tuamotu
Strait

Easter
Island
Strait

DARK-
TROPICAL
SOUTH
AMERICA

Sea
of
Chile

D

WHITE-
TEMPERATE
SOUTH
AMERICA

Drake
Strait

Oceanic straits to
be monitored by Chile

Potential stations
for space travel

Potential paths of invasion from
overpopulated Asian nations
towards South America

Boundary between Dark – Tropical
and White – Temperate South America

Adapted from Riesco (1981, 1984, 1986)

If a general criticism is to be made of the Chilean geopolitical doctrines that insist on that country's dominant role in the Southeastern Pacific, it is that the strategic value given to one of the least trafficked segments of all oceans is exaggerated. Each one of the authors departs from the assumption of a "centrality" of the Southeastern Pacific, which is more fictitious than real. In fact, the articles by Riesco are frequently illustrated by cartographic projections that place Chile or the Antarctic Peninsula in "focal positions" with a great distortion of "peripheral" North America and Eastern Asia. Totally unacknowledged in the Chilean concept is the reality that most of the maritime circulation of the world occurs between Europe and North America in the North Atlantic, or in the North Pacific, where industrialized Japan, Korea, and the Republic of China conduct their trade with North America. It is the very obvious presence of those Asian countries on the Pacific waters, and not artificially built pretenses of centrality, that grants importance— geopolitical and commercial—to the North Pacific and to the countries that border on it. Even assuming that the Southeastern Pacific and the sub-Antarctic waters were heavily trafficked by commercial or military shipping in the case of the destruction of the Panama Canal, the realization of any geopolitical design depends on the capability of the concerned country to maintain a visible presence by means of effective material elements. It happens that Chile, and also Argentina, have only limited naval and air equipment to police the waters of the South Pacific and South Atlantic. The southern outposts and oceanic islands under the sovereignty of Argentina and Chile are practically deserted or only occupied by minute military garrisons. Inadequate ships, land bases isolated from each other, airplanes with restricted cruising range, and lack of modern means to conduct an active occupation and exploitation of the Antarctic reduce all geopolitical programs to mere words and wishful thinking and surround them with a certain air of demagoguery.

In open contrast to the weaknesses of the oceanic geopolitical pretenses of Argentina and Chile stand the geopolitical doctrines of Brazil that emphasize the geostrategic use of the South American mainland. For more than a century Brazil has pursued a policy of continental projection that differs from the oceanic objectives that impelled Argentina and Chile to contend for dominance over the southern tip of the continent and for control of the bioceanic passages.

Already in imperial Brazil, renowned statesmen, among them Mario da Silva Paranhos, Irineu E. de Souza, and Teogilo Ottoni, committed themselves to consolidating internal spaces and to pushing the colonizing frontiers towards the country's peripheries, most particularly into the upper reaches of the Paraguay and Paraná rivers and into the Amazon Basin. Hence, the occupation of the vacant lands contiguous to the Paraná River, the Mato Grosso, and Amazonia enhanced the Brazilian perception that through the consolidation of these interior frontiers the country could project itself effectively into the South American core. Such programs acquired an initial conceptual framework in the geographical writings of Everardo Backheuser. Conversant as he was in German, all later Brazilian geopoliticians owe him their acquaintance with Ratzel, Haushofer, and Kjéllen's geopolitical ideas and with the scores of German geographers who were writing about political and military power as manifested in space at the turn of the century. Indeed, Backheuser's *A estructura política do Brasil* contains most of the fundamental concepts about territory, race, destiny, state, and population that are used by other Brazilian writers in their subsequent geopolitical constructs (Backheuser 1926). Thus, armed with the theoretical premises that were partly imported from Germany by Backheuser and the notions about modern warfare that the French military mission introduced into the country in the 1920s and early 1930s, new figures such as Mario Travassos, Theresina de Castro, Lysias Rodrigues, Golbery do Couto e Silva, and Jaime Ribeiro de Graça, were able to elaborate on the geopolitical role of Brazil in continental South America and the Atlantic Ocean.

It should be noted that by then geopolitical thinking was not merely an intellectual exercise practiced by civilians, but rather a discipline cultivated increasingly by the militaries. Their significance in national programs received a boost through Brazil's participation in World War II on the side of the Allies. Both the involvement of selected Brazilian troops in the campaigns of Italy and a secondary role in policing the waters of the tropical western Atlantic against raiding German warships and submarines enhanced belief in the operational readiness of Brazilian armed forces. Such an opportunity was not granted to their Argentine or Chilean counterparts due to their expressed neutrality in the global conflict. The confidence bestowed by the Allies, particularly by the United States, in Brazilian military skills and their ability to safeguard the Latin American flank of the Atlantic proved to be a powerful incentive to the development of geostrategic doctrines. When, after the war, the Inter-American security zone was instituted under the auspices of the United States, Brazil became the major and most reliable partner in the initiative. The role of policing the western Atlantic introduced a new orientation into the geopolitical program formulated by Brazilian writers. Consolidation of internal frontiers and occupation of thinly populated border regions were no longer the priorities in the geopolitical programs of the country: a destiny on the sea had been discovered and was to support the international action of the country in the years to come, including an increasing eagerness to participate in the exploration and occupation of the Antarctic (Meira Mattos 1979, 78–86).

Obviously, such programs clashed almost totally with the geopolitical plans that the Argentine military had been forging since the rise to power of Colonel Juan Perón. The postwar prosperity of Argentina provided appropriate fuel for the propaganda machine of *justicialismo*, the doctrine that blended elements of nationalism, corporativism, and dreams of grandeur and that became the cornerstone for Argentina's reconstruction. Cited as among the tangible proofs of the nation's potential were its manufacture of automobiles and airplanes. With the cooperation of German aeronautic engineers who had found refuge in Argentina after the war, a jet prototype was built and armored vehicles were developed in the state-owned factories. Impressed with the external appearance of grandeur of the Peronist regime, the Paraguayan dictators Higinio Morinigo and Alfredo Stroessner established dependency relationships with Argentina. No less enthusiastic about the achievements of General Perón was General Héctor Ballivián Rojas, the ruler of Bolivia between 1950 and 1952. Free elections in Chile in 1952 had brought Carlos Ibáñez del Campo to the presidency of the republic. An ousted military ruler who had sought refuge in Argentina, Ibáñez del Campo was a friend of that country and a die-hard militarist who praised the nationalism of Perón and admired the international image of the "new Argentina."

Thus, not by virtue of any geopolitical construct but as a reflection of a regime that appeared to satisfy the longings of the people and the military alike, Argentina found itself the guiding power of the Southern Cone states. Brazil, on the other hand, shattered internally by the chaotic administration of Getulio Vargas, became isolated and was deserted even by its traditional ally, Chile, in its quest to be the leading nation in South America.

The siren calls of Peronism were not heeded by all in Chile. Military officers, Ramón Cañas Montalva in the lead, warned President Ibáñez del Campo about the bellicosity of Argentina and denounced the adventurism of the generals and colonels behind Perón. That indeed an oversized and unruly military establishment had been growing under the cover of populist *justicialismo* surfaced during the second period of Perón's rule, when economic hardship and internal decay revealed the malaise of Peronism. After the ousting of the ruler by a junta representing all armed forces of the country, the military took power in Argentina in 1955 and continued to intervene in politics almost at will until 1982.

Important for our purposes is that the place of international preeminence that Perón and justicialism had acquired in the postwar years was not to be relinquished by the powerful Argentine military; only their methods were to be different. A knowledgeable scholar on Southern Cone geopolitics, Howard Pittman, states that during the 1950s geopolitical thought in Argentina, which had shown some degree of development during the prewar years, "eclipsed," to resurge again in the early 1960s (Pittman 1981). This seems to be substantiated by the prolific production of Argentine geopoliticians in the years after the fall of Perón. One can interpret this resurgence, however, as the search by a powerful military establishment for self-justification and as the endeavor to hide governmental inefficiency behind pompous crusades of national action and international expansion. It is, indeed, symptomatic that every time the socioeconomic conditions of the country deteriorated, the military and scores of civilian followers produced the most diverse and disparate geopolitical programs aimed at reestablishing the leading role that Argentina had enjoyed during the Perón era. Recent history and geographical circumstances offered these thinkers good arguments to back up rosy geopolitical programs, but the motivations were, in most cases, less than legitimate.

Argentina occupies an unquestionable central position in the Southern Cone, and, by virtue of its past, the city of Buenos Aires and the estuary of the La Plata have played a determining role in the political, economic, and cultural affairs of the riverine countries: Paraguay, Uruguay, and eastern Bolivia. Responding to these circumstances, Argentine officials have frequently assumed responsibility for the affairs of the entire Paraná-Paraguay river basin, and this attitude has determined the geopolitical priorities of the country in that area. While Brazil was primarily concerned with the consolidation of its internal frontiers, Argentina had enough time and opportunities to assert itself as the ruling power in the upper La Plata Basin. After that, when Brazil added to its geopolitical programs that of becoming a power in the South Atlantic, Argentina responded to the challenge by revitalizing its maritime vocation, as attested by the writings of Jorge Atencio, Fernando Milia, and Gustavo Cirigliano. Argentina's opening to the South Atlantic, and ultimately into the Antarctic, conflicted with similar programs of Brazil and Chile and set the stage for the confrontation with Great Britain, which had been established on the Falkland Islands since 1833.

From this perspective the mixed emotions of Argentine geopolitical thinkers about being in charge of the destiny of the Southern Cone and, at the same time, encroached upon by hostile nations, become understandable. It is irritating for Argentina that these nations not only obstruct its international programs, but also threaten the integrity of its remote borders. A model of analysis for these frustrations are the axes designed by Gustavo Cirigliano to define the lines of geopolitical action of Argentina, namely, the "fluvial axis" along the Paraná-Paraguay river courses (Map 1.2), which entitles Argentina to participate in international programs with Paraguay, Uruguay, and southern Brazil; the "Andean axis," which borders on Chile and extends Argentina's area of influence into Bolivia and Peru; and the "maritime axis," which legitimizes Argentina's claim for control of the Southwestern Atlantic, including the Falkland/Malvinas Islands, the subantarctic islands of South Georgia, South Sandwich, and South Orkneys. These axes are the main arteries of a "continental Argentina" hinging on the Río de la Plata and the Pampa, a "maritime Argentina" extending between the subantarctic islands, the Falkland/Malvinas Islands, and the Argentine coast, and an "Argentine Antarctic," which includes the segment toward which maritime Argentina is projected (Cirigliano 1975). These constructs are, by no means, the lucubrations of Cirigliano's geographical imagination, but represent a synthesis of the current geopolitical theories that are held by the military, politicians, and diplomats alike.

Map 1.2   Axes and Areas of Argentina's Geopolitical Action

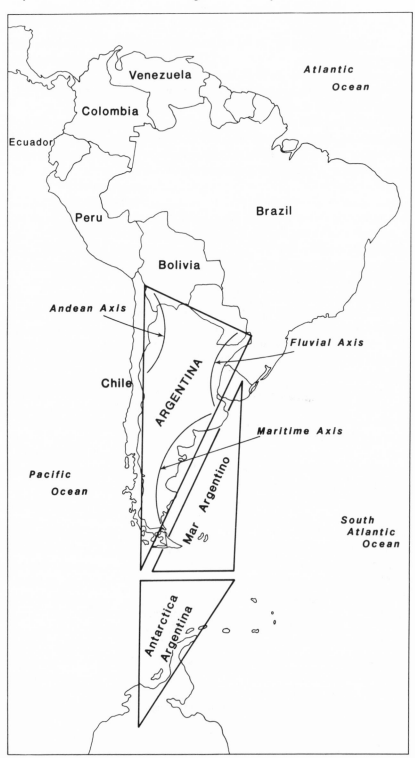

According to Gustavo Cirigliano (1975)

The counterpoint of these widely accepted postulates is that, precisely where the axes of international projection are located, the neighboring countries threaten the territorial integrity of the country through a relentless penetration of migrants across the Argentine boundaries to forestall the Argentine attempts at expanding its sphere of influence. In fact, the coveted "fluvial axis" is presently dominated not only by the economic system of Brazil, but penetrated by land-seeking Brazilians who move into eastern Paraguay (Kohlhepp 1984) and even attempt to acquire land in Misiones. Along the Paraguay River, Paraguayan rural workers enter the Chaco and Corrientes, while others venture into Buenos Aires in search of better opportunities. The northern edge of the Andean axis is flooded by Bolivian laborers seeking employment in Salta, Jujuy, Tucumán, and Mendoza; in its southern edge the penetration of Chilean rural workers is evident in the fields of Río Negro, Chubut, and Patagonia. The presence of thousands of Chileans in Argentine Patagonia and Tierra del Fuego has become a particular concern for Argentine authorities, who consider them a potentially dangerous "fifth column" in a hypothetical case of armed conflict with Chile (Mármora 1978). This perceived presence of "the Chilean state" in territories scarcely populated by Argentines is regarded as a threat for the continuity of maritime Argentina and a hindrance for the projection of the country into the Antarctic. It is in this context that we must view the intransigence with which Argentina pressed Chile to drop its claims for full control of the Beagle Channel and to use Cape Horn as the focal point of Chile's projection into the Southern Ocean and the Antarctic. That this area is, indeed, a most sensitive spot in the geopolitical postulates of both Argentina and Chile is shown by the decades of unsuccessful Chilean attempts to establish settlers on the shores of the Beagle Channel and to populate the inhospitable Picton, Lennox, and Nueva islands on the eastern mouth of the channel, and by Argentina's active promotion of the creation of a permanent industrial center in the town of Ushuaia in order to attract its own people to the Beagle Channel (F.I.D.E. 1983).

Of course, these attempts by the two countries to create artificial population centers in order to ascertain the principle of sovereignty may appear unrealistic and costly, but they are imposed by the geopolitical dictate that the nation with the most effective control of the southern tip of South America will have priority access to the Antarctic.

GEOPOLITICS OF MISTRUST OR POLITICS OF COOPERATION?

Geopolitical constructs persist and enjoy a relatively high degree of credibility, even among the well-informed and educated populations of the Southern Cone. The reason for their persistence is that the issues involved are wrapped in a cover of historical scientism and geographical pertinence that appeals to national sentiments. For nationalists of all shades there is nothing more comforting than to believe that pressing priorities to improve the social and political conditions can be replaced by the mobilization of a nation behind nationalistic causes. Far too often, however, dreams or national frustrations have become dangerous incentives to embark upon militaristic adventures, making a country rally behind the programs of the military.

In the framework of these programs it is common that the national or international significance of certain issues pertaining to the relations between neighboring states are distorted or exaggerated by the geopolitical imperatives dictated by the proponents of such doctrines, with total disregard for the common good of the nation. Moreover, there is the problem of latent irredentism. For as long as the feeling exists that certain territories, lost in times of adversity or military weakness can be regained by means of military actions, the potential for friction and animosity will persist. Thus, division and competition rather than union and cooperation have been promoted by most geopolitical doctrines current in Southern Cone countries.

Only recently has the strictly geostrategic or national-security character of geopolitical doctrines been replaced by the idea of the geopolitics of economic integration. In this concept, the spheres of action of a particular state are determined not by military prowess but, instead, by the integration of adjacent countries into the commercial networks of the most advanced states. Financial strength, technological superiority, and capital surplus determine the domination of one nation over the less advanced ones. Underdeveloped countries are regarded as sources of inexpensive labor or natural resources and as outlets for goods produced by the technologically advanced nations. Trade is ruled by the state with greater production capacity and the most dominant currency. Within this perspective, the advanced states promote the formation of "integrated economic regions" that are to be articulated along vital arteries of communication, such as waterways and international roads. From nodes along these arteries of communication, new economic valorizations in regions of scarce development or into unexploited areas are undertaken, applying the technological know-how and the financial muscle of the best developed partner. In these schemes, the *resource frontiers* of the past are stimulated and new exploitation fronts are sought for future incorporation. Cooperation and complementation become the main features and binding forces of these programs. Thus arose the "Río de la Plata Basin Initiative" to integrate the economies and development plans of the countries with rivers emptying into the La Plata estuary, and the "Amazon Pact" to structure the economies of the countries with segments of their territories in the Amazon basin (Meira Mattos 1977).

Well-intentioned as these initiatives appear on paper, they cannot be disassociated from Argentina's determination to dictate further the course of affairs in the La Plata Basin and from Brazil's determination to monopolize the development of the Amazon Basin. It is precisely because of these unabandoned agendas that the minor partners' adherence to both initiatives has been less than enthusiastic and their implementation extremely sluggish. Despite these limitations, future geopolitical models will probably insist on the creation of these spheres of interest instead of dwelling on the traditional critical areas that were determined by geostrategic considerations.

This new accent on the economic valorization of geographical space and the monopoly of unexploited resources also explains the insistence of Argentina and Chile on gaining control of the Southern Ocean and, ultimately, of the South American quadrant of the Antarctic. Not only the mineral resources of that continent, but also the fish and krill of the subantarctic waters have increasingly become a bone of contention between Argentina and Chile, with Japan—the relentless predator of the world's oceans—Taiwan, the Soviet Union, and South Korea all fiercely competing for the region's ocean bounty (Knox 1983). We must consider in this context, therefore, the strong endorsement made by Argentina and Chile of the "territorial sea" or "patrimonial sea" postulates during the last conference on the Law of the Sea. It is interesting, however, that Argentine and Chilean administrators realized that their independent declarations of sovereignty and exclusive use of the resources of the seas and lands of the Antarctic could not be upheld without a strong military stance against the advanced countries that had crept into their backyards in order to exploit them. Thus, although Argentina and Chile still have converging interests in the Antarctic, both countries have lately opted for collaboration in rejecting the intrusion by distant nations into territories and waters traditionally regarded as theirs. The common front they presented on Antarctic Treaty issues is the first proof that cooperation is possible among nations of converging—and at times colliding—geopolitical interests (Selcher 1985, 30–33; White 1986, 74–78).

This by no means signals the demise of some geopolitical doctrines that have been supported with vehemence by the countries in question. Over the years these doctrines have become so deeply ingrained in the idiosyncracies of the peoples of the Southern Cone nations that relinquishing them would be considered as betrayal of the nation, abandonment of its glorious traditions, and repression of cherished dreams of grandeur still latent in all these countries. Marriages of convenience, such as those mentioned above, will occur around certain issues of mutual interest, but in the foreseeable future as long as there exist military establishments ready to justify their existence, competition, and desires "to be the first," any plans for long-lasting and constructive cooperation will be doomed.

chapter two

Harmony or Discord: The Impact of Democratization on Geopolitics and Conflict in the Southern Cone

Howard T. Pittman

Academic and media analysts have originated considerable commentary on the effects of democratization in Latin America since the return of democracy in a majority of those countries. In the Southern Cone, Argentina, Bolivia, Brazil, Peru, and Uruguay now have democratic governments. Only Chile and Paraguay remain under authoritarian rule. Most U.S. scholars seem to predict a new and favorable era in Latin American politics and international relations based on the democratic trend, despite continuing economic and social problems facing these nations. Latin American scholars and government officials echo these forecasts and frequently attribute all good events or trends to the return to democracy. In May 1986, for example, the president of Venezuela acclaimed a local beauty queen as "the product of democracy."

The question remains whether these optimistic forecasts are valid in the face of reality, that is, the ongoing and sometimes increasing economic and social problems endemic to the region and the new and old disputes and controversies which plague regional international relations. This essay examines the impact of democratization on geopolitics and conflict, including the effects on new and continuing continental and hemispheric disputes as well as pertinent offshore conflicts involving the Southern Cone nations.

GEOPOLITICS AND CONFLICT IN THE SOUTHERN CONE

Actual conflict began with the first colonial settlements struggling against the Indians, which was solved relatively early by means of slavery and miscegenation in Brazil, but which continued for three and a half centuries in Argentina and Chile, delaying the occupation of Patagonia until the latter half of the past century. There was also strife between the colonies of Portugal and Spain, especially in the La Plata Basin, where Portuguese explorers probed and established settlements on the north bank of the Plata and west to the Paraná, far beyond the Tordesillas Line, bringing them into contact with Spanish colonies in struggles against British, French, and Dutch invaders who refused to accept the Papal division of the world between Portugal and Spain. French and British intervention continued into the middle of the past century. Spain attempted to recover some of her past colonies in the 1860s. Britain still retains the Malvinas and other South Atlantic islands after fighting off an Argentine takeover in 1982.

After Independence, conflict and warfare continued for many years. Massive transfers of territory occurred. Brazil, continuing her expansion until the early 1900s, gained territory at the expense of most of her neighbors; Argentina at the expense of Bolivia, Chile, and Paraguay. Major wars were fought over Uruguay and Paraguay. Chile broke up the Bolivia-Peru federation and later, in the War of the Pacific, gained territory from both Bolivia and Peru. The last major war, the Chaco War, took place in the 1930s.

The record of historical conflict includes territorial disputes dating back to the age of discovery and postindependence expansion of some states at the expense of others. The continuing nature of these disputes still influences contemporary regional relations. Arguments concerning the Tordesillas Line, the 1750 Treaty of Madrid, and the decrees, regulations, and maps of the Spanish empire are still being cited in such conflicts as the Beagle Channel dispute, the Malvinas War and Antarctic claims. The 1982 Malvinas War represents the first armed strife with a European power in nearly a century. Ironically, while there are no active territorial disputes between Brazil and her Hispanic neighbors today, there are a number of continuing conflicts among the Hispanic American states.

The historical memory is long in South America: disputes dating back to the age of discovery remain active today. Beginning in the 1940s, with the establishment of Antarctic claims by Argentina and Chile, disputes among countries of the region have been extended into offshore areas, including maritime space, islands, and Antarctica, broadening the areas of possible conflict to vast portions of the South Atlantic, South Pacific, and Antarctic oceans and, in the case of the Malvinas War, once again involving extracontinental powers.

Geopolitical ideas and actions in the history of the Southern Cone parallel the conflict. Luso-Brazilian expansionism, legitimized through the ideas and efforts of Alexandre de Gusmão and the Baron of Rio Branco and the concepts of José Bonifacio are frequently cited (Meira Mattos, 1977:41-68) as geopolitical in nature. The attempts of the kings of Spain to colonize the Strait of Magellan, the establishment of a separate Captain Generalcy in Chile, and the creation of the Viceroyalty of the Río de la Plata have been identified by an Argentine scholar (Marini 1986a, 7–26) as geopolitical, as have the postindependence ideas and policies of O'Higgins and Portales in

Chile and those of San Martín, Mitre, Sarmiento, and Roca in Argentina.

Contemporary geopolitical theory in the Southern Cone based on the historical record and national goals and aspirations provides the rationale for existing continental disputes as well as their extension into offshore areas. The influence of geopolitics upon international relations in the Southern Cone is the result of the emergence of a pattern in the development of indigenous geopolitical theory and its application to government plans, policies, and actions, particularly in the ABC countries (Argentina, Brazil, and Chile) but also to some degree to Bolivia, Uruguay, and, to a lesser extent, Peru. This pattern is discussed in detail in Pittman (1981, 395–498, 835–912, 1339–1413).

The relationship between geopolitics and conflict appears throughout the Southern Cone and now extends to offshore areas where the nations concerned contend for resources, territory, and influence. There are three basic types of conflict evident today. First, there are existing historical territorial and border disputes. Second, there are disputes over *vacíos* (underpopulated, underdeveloped areas within a given country), which, it is feared, may be lost through foreign migration and investment. Third, and most important, are national claims to offshore islands, Antarctica, and the sea itself. Economic gain, the needs for development and the growing competition for these, especially energy, mineral, and food resources, affect all three categories of disputes, particularly those involving offshore claims.

It is these conflicting offshore claims that imply future conflict both among regional powers and between those powers and extra-regional claimants. A process is in motion whereby geopolitical concepts once applied only to land space are being extended offshore to new conflicts over control, possession, and integration of these areas and their resources. In this process, concepts of the sea as sovereign territory, and the equivalence of land and sea space are beginning to be used in national policies and international negotiations. In the most extreme cases of geopolitical thinking, island and Antarctic territorial claims are now linked to mainland national territories through nationally claimed, controlled, and demarcated seas which extend far beyond the 200-nautical mile (nm) economic exclusive zone, creating new areas of conflict and increasing the probability of future conflict with competing extraregional powers.

THE BOLIVIA-CHILE-PERU TERRITORIAL DISPUTE

This conflict dates back to the past century, when Chile, after helping Peru secure freedom from Spain, first broke up a Bolivia-Peru federation and then, in the War of the Pacific, extracted resource-rich territorial concessions from both nations, leaving Bolivia landlocked. Pittman (1981, 1054–1062), after a detailed analysis, argued that this was a classic case of geopolitical action. In a more recent study, Morales (1984, 171) stated that:

> Geopolitics as the central determinant of Bolivian foreign policy captures the essence of this country's past, present and future; for in few countries has geopolitics been as important and constant as is the Bolivian struggle for political and economic sovereignty.

The landlocked status of Bolivia since 1879 has made the central goal of her foreign policy to be the *salida al mar*. Chile and Peru finally arrived at a settlement in 1929 which gave Tacna to Peru and Arica to Chile and provided that Chile could not transfer any part of the former Peruvian territory to a third party without Peru's consent, thus creating a two-party barrier to Bolivian access to the sea. Bolivia has launched repeated campaigns since 1879 to secure an outlet to the sea without success. It is curious that while Bolivia has lost more territory to Peru (92,527 square mile) than to Chile (46,223 square mile), it has always been Chile who is the target of Bolivian demands for a sea outlet.

The last serious Bolivian-Chilean negotiations took place from 1975 to 1978 between the military governments of Pinochet and Banzer. These negotiations introduced the concept of the equivalence of land and sea space, that is, that both land and sea are sovereign space and can be of equal value and subject to transfer from one country to another by agreement. The original offer proposed granting Bolivia a corridor along the border of Peru extending into the sea to the 200-nm limit, in exchange for an equivalent amount of Bolivian land territory equal in size to the total area of Chilean land and sea space in the proposed corridor. The offer was initially accepted by Bolivia, which did, however, object to the extension to the 200-nm limit, but not to the patrimonial sea. The land portion of the proposed corridor was located in the territory covered by the 1929 agreement, and its transfer was subject to Peruvian concurrence. In 1976, Peru presented her proposal for a much larger Bolivian access to the sea through Chilean territory, ending in a tripartite coastal enclave north of the city of Arica, in which sovereignty would be shared by all three nations. The Chilean port of Arica would be administered by a trinational port authority, while Bolivia could develop her own port in the shared tripartite zone. Chile rejected the Peruvian proposal as "an infringement of Chilean sovereignty." Bolivia welcomed the idea but broke off negotiations in 1978.

Since that time, Bolivian governments have continued to seek support for their claims and to condemn Chile in every possible international forum. They have also supported Argentina in the Beagle Channel and Malvinas disputes in return for Argentine backing for their claims (Pittman 1981, 1469–1475; Morales 1984, 181, 191). The salient feature of the Bolivian claims is that they always start with the premise that the world and particularly Chile owes Bolivia a corridor to the sea; thus the only thing to be negotiated is the size of the concession and the timetable for its surrender, an attitude that also characterizes the Argentine claims to the Malvinas, as will be shown later in this study. Bolivia has also sought to take advantage of the unpopularity of the Chilean authoritarian regime to pursue its claims, an example of Grabendorff's (1982) "system conflict." Abortive talks with Chile in Santiago in May 1987 were broken off because of alleged Chilean intransigence.

It is pertinent to our analysis that the last serious negotiations took place when all three countries were under authoritarian regimes. These were no more productive than the previous initiatives of socialists Allende and Torres (Morales 1984, 181). A democratic government in Peru has not changed its attitude, although some abortive military staff talks, apparently unrelated to the territorial dispute, took place with Chile in 1986.

What is evident is that this old dispute, now involving new concepts of the equivalence of land and sea space, remains unsolved. It provides an example of a recourse to system conflict that has proved just as unsuccessful as previous negotiations between differing types of regimes. Moreover, in the light of history, it is unlikely that the accession of a democratic government in Chile will automatically result in the settling of the dispute, since past governments of all types have supported the nation's geopolitical goals. The impact of democratization on this dispute must be regarded as nil.

SOVEREIGNTY OVER SEA SPACE:
A NEW ARENA OF CONFLICT

The trend toward extending sovereignty into the sea was apparently triggered by the U.S. declaration regarding the continental shelf, which was seized upon by South American countries for their own purposes, and has been accelerated by the concept of a 200-nm economic exclusive zone, first popularized in South America. Within this new trend, geopolitical thinking in the Southern Cone already views the maritime space within the 200-nm limit as sovereign territory. At the same time, "national seas" of much greater scope, linking island possessions and Antarctic claims with the mainland, are being conceptualized, advocated, and sometimes proclaimed. The idea of sovereignty over maritime space seems to have originated in Chile, always a maritime nation, but has been accepted in Argentina, Uruguay, and, to a lesser extent, in Brazil. This concept originated in the past century, but the idea has gained recent popularity as a result of resource competition. Disputes over water areas date back to 1843, when Chile occupied the Strait of Magellan, despite Argentine protests. As part of the settlement of the dispute over Patagonia, the countries agreed that Argentina would control the Atlantic access, and Chile the Pacific, but both nations have made repeated attempts to overturn or modify this principle. In the 1950s Argentina introduced the idea of the Diego Ramírez meridian as a divisor, which would have moved the line several degrees westward into the Pacific. Chile countered with the concept that the Pacific extended eastward into the Arc of the South Antilles, which would have shifted the divisor eastward into the Atlantic, cutting off Argentine access to Antarctica. Argentina returned to the concept of the Cape Horn meridian as the Atlantic-Pacific divisor, a position maintained to this day, particularly during the Beagle Channel controversy.

The Concept of the National Sea

The concept of a national sea was first developed in Chile. The 1940 Chilean Declaration of an Antarctic Territory included "the respective territorial sea" within the limits of the sector between fifty-three and ninety degrees west longitude. Introduced in the 1950s was the concept of a "Mar Chileno" (Ihl C. 1951), bounded by lines drawn from the Chile-Peru border westward to the Nazca Ridge and the Easter Islands and thence southward to the western boundary of Chile's Antarctic claim. In 1976 the Chilean Sea was defined in the Chilean national Maritime Policy (Orrego Vicuña 1977), although no boundaries were established. The concept was updated to include marking the maritime boundaries with buoy markers (Marull Bermúdez 1975). Meneses (1981) would fortify the islands, using them as air and naval bases to control the Chilean Sea, which would also be marked by floating bases and sonobuoys to detect submarine as well as air and surface movements. While not necessarily related, the airfield at Easter Island was upgraded in 1987 at U.S. cost to provide an emergency landing site for the space program. This also increased its value as a Chilean defense base capable of supporting the largest jet aircraft.

Both the Argentine and Brazilian declarations regarding the territorial sea emphasize sovereignty. The Argentine Decree Law No. 17094 of December 1966 established a 200-nm limit, stating in Article 2:

> The sovereignty of the Argentine nation extends in the same way to the seabed and the subsoil of the adjacent submarine zones to a depth of 200 meters or beyond these limits as far as the depth of the waters permits.

Based on this concept, Villegas (1982) argues, the Malvinas, Georgias, and South Sandwich islands are a continuation of both the continental platform and the national submerged territory and, further, that they are the keys to the control of the Strait of Magellan, the Beagle Channel, and the Drake Passage. Thus, the nation who controls these passages can exercise control of maritime traffic between the Atlantic and Pacific and the continent and Antarctica as well. This observation is shared by Chilean geopolitical analysts who oppose the return of the Malvinas to Argentina for this reason.

Similarly, Brazil's law on the territorial sea (Decree Law 1098 of March 1970) established a belt 200 nm wide starting from the tide mark of the continental and insular coasts. Article 2 of that law states that "the sovereignty of Brazil extends in the air space above the territorial sea as well as to the seabed and subsoil of this sea."

Sea-Boundary Precedent

A precedent for extending boundaries to the 200-nm limit was set in 1973 by the *Tratado del Río de la Plata y su Frente Marítimo* (Treaty of the Rio de la Plata and its Maritime Front) between Argentina and Uruguay, which established a maritime boundary between the two nations out to the 200-nm limit at the mouth of the Plata estuary. This precedent, plus the experience gained in the Beagle Channel and Malvinas disputes, have led to the concept of a large "Mar Argentino" similar in size and scope to the Chilean sea. This Argentine Sea (Boggio Marzet 1982) would extend from the Argentine-Uruguayan sea boundary eastward to its intersection with the meridian 21 degrees, 27 minutes west longitude (Point Admiral Brown), thence south to the pole and bounded on the west by the edge of Argentina's Antarctic claim and the Cape Horn meridian. The western boundary of this sea would conflict with the Chilean sea where the two countries' Antarctic claims overlap.

National Sovereignty, Sea Space, and Conflict

The concept of national sovereignty over maritime space has not yet in itself provoked armed conflict over maritime boundaries, but it did influence claims and negotiations in the Beagle Channel case and sea battles in the Malvinas War. The latter, if one believes Argentine sources, involved sea as well as land goals. Some of the conflicting geopolitical thinking in the Southern Cone could lead to future combat, not only among these states, but also between them and extracontinental powers. The impact of democratization is still speculative in this arena, except in relation to other conflicts, in which claimed sovereignty over the territorial sea may very well extend and exacerbate future disputes, regardless of the ideological bent of the governments concerned.

THE BEAGLE CHANNEL DISPUTE

The concept of maritime space as territory intensified the long-standing Beagle Channel dispute between Argentina and Chile, which almost led to war at Christmas of 1978. The dispute, dating back to the turn of the century, is too long to be reviewed in detail here; for more extensive accounts see Pittman (1981, 1986), Estrategia (1978), and Andreasen (1978). After years of contention, the dispute was placed in arbitration in 1971; an award made in 1977 established a boundary in the Beagle Channel and gave Chile the islands south of it. Chile accepted the settlement; Argentina refused to do so and declared the award null, then attempted to achieve its goals through bilateral negotiations and threats of war until the end of 1978, when papal mediation began. A papal proposal in 1980 was resisted by Argentina, but a settlement was reached in the 1984 Treaty of Peace and Friendship.

The reasons behind the dispute were both strategic and economic; the real issue was control of maritime space under the new concepts of sovereignty over the sea. This dispute was not about the sovereignty of the channel islands, which had been under Chilean administration for almost a century, but about control of maritime space now considered as national territory. Chile claimed that the award, together with the 200-nm limit based thereon, gave it control of maritime space. In the Argentine view this not only deprived Argentina of the use of this portion of the sea, but also cut across lines of communication to Argentina's Antarctic claim. Basing its actions on the 1977 award, Chile established *líneas de base rectas* (straight base lines) enclosing the islands granted by the settlement and appointed *Alcaldes de Mar* for them, acts of sovereignty which aroused Argentine protests and reactions.

Argentine reactions and Chilean responses were both couched in terms of sovereignty. In rejecting the 1977 award, the Argentine foreign minister stated (Montes 1978):

Map 2.1   The Beagle Channel and Tierra del Fuego Area

The Argentine government reaffirms sovereignty over the insular territories, maritime spaces and continental platforms and seabed that are located in the Atlantic Ocean east of Cape Horn.

The Chilean foreign minister referred to the "clear rights and undoubted sovereignty of my country in the territories and maritime zones of the southern region," while President Pinochet later argued that the principal divergence resulted from "the necessity to delimit the maritime jurisdiction of our states in the southeastern zone of the continent," emphasizing that submarine areas could be considered as part of the territory over which the state has dominion, "that these areas, although covered with water, are a prolongation of this territory beneath the sea."

The Tratado de Paz y Amistad
(Treaty of Peace and Friendship)

This treaty of October 19, 1984, established the maritime boundaries that had been the subject of years of negotiations and mediation. Article 7 describes the delimitation line in detail in the Beagle Channel and vicinity, while Article 10 establishes a maritime boundary at the eastern mouth of the Strait of Magellan. The effect of these provisions is to prevent Chile from projecting a 200-nm economic exclusive zone into the Atlantic. Article 11 confirms the *líneas de base rectas* established by both nations. Annex 1 provides for consultation and arbitration of future disputes and Annex 2 prescribes strict rules of navigation for Argentine, Chilean, and third party ships in the Beagle Channel, the Strait of Magellan, and the Lemaire Strait, thus assuring Argentine access to the base at Ushuaia and Chilean access to the Atlantic through the Lemaire strait.

What this treaty does is to: (1) confirm the 1977 award which demarcated the channel and gave the islands south of it to Chile; (2) establish maritime boundaries in the long disputed sea zone, which prevents Chile from claiming economic exclusive zones in the Atlantic; and (3) confirm the Cape Horn meridian (67 degrees, 15 minutes west longitude) as the divisor between the Atlantic and the Pacific. The treaty also provides that its arbitration and conciliation articles (1 through 6) will also be applicable to Antarctica, but with the proviso (Article 15) that this in no way can be interpreted to affect the rights, sovereignty, juridical positions, or delimitations of the parties in "Antarctica, or its adjacent maritime space, including the seabed and subsoil." Thus, the treaty clearly links the territorial disputes at the tip of the continent with those in Antarctica and surrounding waters, and it confirms the relation of respective geopolitical goals of Argentina and Chile to the provisions of the treaty.

Analysis

What was the impact of democratization upon the resolution of the Beagle Channel dispute? Does the 1984 treaty represent a final settlement of the matter? A majority of Argentines (82 percent) approved the treaty but hardline Argentine sources (Rattenbach 1984; Diaz Besson 1984) argue that the treaty was arrived at too soon and believe that future conflict is likely to occur, a position also held by a Chilean analyst (Santis 1984). An Argentine editor (Bravo 1984), on the other hand, advocates using the treaty as the basis for physical integration, scientific and cultural cooperation, bilateral development, and Latin American unity.

If there had been a democratic government in Chile, would the results have been different? The military government of Chile made no military threats, nor engaged in actions to force a solution, and Argentina could have had a settlement much like the 1984 treaty as early as 1978. Some Argentines (Russell 1984; Zollezi 1984) questioned the wisdom of making an agreement with a discredited military regime in Chile, while the Chilean Christian Democratic party was opposed to negotiations with Argentina. The efforts to settle the dispute were begun by presidents Allende and Lanusse, continued through negotiations between two military governments from 1977 to 1982 and finally resulted in a treaty between an elected democratic regime in Argentina and an authoritarian government in Chile. While the rise of a democratic government in Argentina was a factor in reaching an accord, neither nation has abandoned its geopolitical goals. One might also conclude that the Alfonsín government, having severely weakened an already defeated armed forces, had no choice but to conclude a treaty with its old rival in the face of its weakness and the continuing dispute with Britain over the Malvinas islands. The impact of democratization is evaluated as a contributing but not the only operative factor in the 1984 settlement.

THE MALVINAS DISPUTE

The Malvinas/Falklands dispute dates back to the colonial era, when Great Britain and Spain claimed rights of discovery and attempted to colonize the islands. The roots of the present controversy lie in the 1833 British reestablishment of a colony there, an act Argentina has protested ever since. The conflict led to war in 1982, with Britain regaining control of the islands over which Argentina still claims sovereignty. For detailed reviews of the Malvinas dispute, see Child (1985), Diaz Bessone (1982), Gamba (1987), Moro (1985), and Pittman (1981, 1986).

In the climate of decolonization after World War II, Argentina renewed her claim to the islands in the United Nations and other international forums. Several negotiations were held without success. By Decree Law 2191 of February 25, 1950, Argentina incorporated the islands into the Argentine

National Territory of Tierra del Fuego, Antarctica, and the Islands of the South Atlantic included between the meridians of 25 and 74 degrees west longitude. On April 2, 1982, unsatisfied with British responses, Argentina seized the islands; Britain retaliated with an amphibious campaign and recaptured the Malvinas on June 14, 1982. The Argentine government fell and was replaced with a new one, but both old and new regimes stated that Argentina would not give up the struggle for the Malvinas (Diehl June 23, July 2, 1982).

The coming to power of a democratic government in Argentina did not change the Argentine position nor result in the cessation of the claim, while Britain has expanded the airfield, developed more base facilities, and continues to patrol the surrounding seas. In 1983 Argentina charged that the British military buildup threatened Latin American military security, an accusation denied by the British foreign secretary Howe, who stated, however, that Britain "will defend the Falkland Islands and their inhabitants against the possibility of renewed attack."

In 1984, negotiations with the British broke down over Argentine insistence that the sovereignty of the islands be included on the agenda. President Alfonsín reiterated that any solution to the problem "must be based on the recognition that the Malvinas are, were, and will be, Argentine," a position he has maintained ever since. Swiss efforts to get the two parties back to the negotiating table failed, according to Swiss foreign secretary Brunner, because "the problem is that one country wants to talk about everything but sovereignty and the other country wants to talk about nothing but sovereignty."

In the interim, nothing has really changed in the position of the two parties. In the Malvinas/Falklands dispute, the position of the "democratic" government of Argentina remains exactly that of its authoritarian predecessors, to wit: Argentine sovereignty is not negotiable. In other words, Britain must recognize Argentine sovereignty before negotiations can proceed, which would be limited to the phase-out of British rule, an attitude not conducive to peaceful solution of the dispute.

The Malvinas conflict illustrates some of the same issues affecting other historical conflicts and, at the same time, reintroduces the involvement of an extracontinental power, a situation that could result in additional conflict in areas claimed by Southern Cone countries. The struggle is historically based; the contested islands, while located close to Argentina, are populated by British colonists who are neither Argentine nor Latin. This enclave of British subjects adds a dimension of extra-continental involvement that the Argentine state traditionally has opposed. The Argentine claims have gained importance under new geopolitical concepts of sovereignty over sea space. They are now supported by extended claims to maritime space, other South Atlantic islands, and Antarctica, which overlap British claims to some of the same territories. The impact of democratization on the Malvinas case must be regarded as negative, since the democratic government of Argentina has changed neither the geopolitical goals nor the position of its authoritarian predecessors.

CONFLICTING CLAIMS TO ANTARCTICA

As noted in the sections above, both the Beagle Channel and Malvinas islands controversies are inextricably linked with the national claims and geopolitical thinking of Southern Cone nations in regard to Antarctica. Conflicting claims involve both land and sea areas. The principal rivalry is between Argentina and Chile, but new interest has been aroused in Brazil and Uruguay, and one international school of geopolitical thought would exclude all extracontinental powers, starting with Britain, from a so-called South American Sector defined as extending from 0 degrees to 120 or 150 degrees west longitude. Space limitations preclude a full discussion of the history, geopolitical and strategic goals, and contemporary actions of the nations involved. For more information see Child (1988), for the most comprehensive study by a U.S. scholar; Fraga (1978) for Argentine views; Pinochet de la Barra (1944) for Chilean positions; and de Castro (1976) and Menezes (1982) for Brazilian concepts.

Argentine and Chilean claims are based on the *uti possidetis* of 1810 as the heirs of Spain, while Brazilian concepts rest on adaptation of an Arctic theory that has been applied to the Antarctic. Both Argentina and Chile, as well as Great Britain, began attempts at exercising sovereignty around the turn of the century. In the 1940s, exploration and scientific activities by other countries, especially Britain, resulted in formal claims to Antarctic territories. The Chilean Decree Law 1747 of November 6, 1940, forecast the present trend toward viewing maritime space as territory and including it in a larger claim:

> The Chilean Antarctic, or Antarctic Territory, is formed by all of the lands, islands, islets, reefs, and glaciers (pack ice) known and others to be discovered and the respective territorial sea between the limits of the sector formed by the meridians of 50 and 90 degree west longitude.

Both Chile and Argentina began establishing bases in Antarctica in the same decade and have continued various sovereignty-building measures ever since. Although both nations signed the Antarctic Treaty, they have continued to maintain their respective bases and claims, and their national maps show their own Antarctic Territories. The growing scarcity of resources worldwide has led to new interest in the exploitation of resources in Antarctica, both on land and sea, an interest not limited to South American states but evident as well among Third World nations intent on securing a portion for their own use and benefit.

Strategic considerations are also important in this conflict. Bases on both continents would be necessary to control the Drake Passage, which is becoming increasingly important in the days of superships and of the possible closing of the Panama Canal. Control of the Drake Passage and the Strait of Magellan makes possible the interdiction of both north-south communications between South America and Antarctica and between the Atlantic and Pacific oceans as well, a fact well known to theorists of the ABC countries (Fraga 1978; Menezes 1982; and Cañas Montalva 1953). This accounts in part for the Beagle Channel and Malvinas disputes, as noted above, and it is intensified in the conflicting claims to Antarctica. Recent Argentine concepts regarding creation of an extensive Argentine Sea in the South Atlantic–Antarctic oceans (Milia et al 1978; Boggio-Marzet 1982) overlap the dimensions of the Chilean Sea, and these potential disputes, in addition to new Brazilian interest in the southern straits area, are causing regional and international concern.

The Brazilian theory of *defrontação*, or "facing theory," argues that all South American countries whose coasts face Antarctica should be entitled to a claim therein based on their respective land boundaries and island possessions (de Castro 1956, 1976). This would not only create a claim for Brazil, based on the meridians of Martim Vaz and Arroyo Chui (which would conflict with Argentine, though not Chilean claims), but would also establish a small claim for Uruguay, as well as claims for Peru and Ecuador (which would conflict with that of Chile). It would also reserve a wide South American Sector which would conflict with the existing claims of some European nations and, by implication, eliminate the idea of a world heritage of mankind as advocated by the Third World nations. Brazil has adhered to the Antarctic Treaty and, without making a claim, has declared "a direct and substantial interest in Antarctica," pointing out that under the Inter-American Treaty of Reciprocal Assistance Brazil is also responsible for the defense of Antarctica. In addition, Brazil has established a National Commission for Antarctica and a Brazilian Antarctic Project to oversee Brazilian expeditions and a new base in Antarctica.

Some of the Brazilian concepts have been complemented by Argentine and Uruguayan geopolitical thinking. Argentina has long advocated the creation through integration of a monolithic Latin American bloc that could deal on a more equal basis with the superpowers and other regional power structures. Argentine geopolitical theorists are now talking about the formation of a continental bloc for the exploitation of South Atlantic and Antarctic resources. Palermo (1980) urges the creation of a Southern Cone bloc led by Argentina and Brazil, including Uruguay, Paraguay, and Bolivia, with a bloc strategy for the exploitation of South Atlantic–Antarctic resources. Admiral Milia's concept of Atlantártida (1978) and Boggio-Marzet's concept of an expanded Argentine Sea bolster this idea. A Uruguayan analyst, Crawford (1982), recommends the creation of an "Ibero-American Club" before the review of the Antarctic Treaty in 1991, which would declare the resources of the area between 25 and 150 degrees west longitude as the "common heritage of America." This would include existing and possible future Latin claims under the theory of *defrontaçáo*, and would exclude claims of non-Latin American powers (such as Great Britain), as well as the concept of Antarctica as the "common heritage of mankind." The concept of a concerted South American effort regarding Antarctica and other claimed and possessed resources received support in 1984 from the military defense committee of *Unidad Popular Argentina Latinamericana*, an association of geopolitical thinkers with representatives from most South American countries. The committee adopted the following proposals pertinent to the issue:

- To assure that the exploitation of their own natural resources render the maximum benefits to the people of the Latin American countries.
- To champion the sovereignty of the Latin American Antarctic Sector as a reserve for future exploitation, supporting the formation of a common strategy and joint operations in the zone as a way of preserving it from the interests of imperialist dominion.

Do these ideas and proposals bear any relation to reality? If so, what are the implications for the future? Personal research, supported by news reports (Anderson 1982), leads me to the belief that the Malvinas War was but the first step in a process of eliminating the claims of non-Latin powers in the area, starting with what was seen as a declining European power, Britain. The Southern Cone concepts of sovereign maritime space, the maintenance of existing claims and the potential for more Ibero-American claims, new initiatives toward exclusive exploitation of Antarctic resources, and the frequent references of nationalistic scholars to the approaching review of the Antarctic Treaty in 1991 all could lead one to conclude that the contest over Antarctica and its resources may well intensify in the decades ahead.

There seem to be two different concepts regarding the future of Antarctic claims in the Southern Cone. One, a hardline approach, would maintain present national claims, goals, and activities against all other claimants. The other seeks to achieve some type of Southern Cone or South American grouping or "club" dedicated to preserving Antarctica and its resources, including those of its surrounding waters, as the "common heritage of America," rather than that of mankind, as advocated by nonregional Third World states. There is strong opposition to British claims and a growing consensus to try to limit or constrain U.S. policy in Latin America in all fields. Finally, the future action of other claimants and the interests of the superpowers must be considered. Other analysts, such as Jack Child (1985), recognize the implications of the problem. Child notes that:

Regardless of the eventual outcome, be it a continuation of the treaty regime, internationalization, or a "new Antarctic Order" with a South

American sector, geopolitical thinking in the Southern Cone will be increasingly important as we move toward 1991.

The impact of democratization on conflicting Antarctic claims cannot be evaluated at the present time. As Child (1987, 177) points out, if the impact of geopolitical thinking can be constructive, emphasizing cooperation and integration, it can also be dangerous. How the concerned nations proceed will determine the future. There are no indications of their abandonment of geopolitical goals. When and if they grow stronger and the concept of reserving Antarctic resources for South American use gains wider acceptance, the possibility of future conflict cannot be ignored, whether or not the governments at the time are authoritarian or democratic in nature, especially since the democratic governments now in power show few, if any, signs of giving up their geopolitical goals.

IMPACT OF DEMOCRATIZATION ON U.S.–LATIN AMERICAN RELATIONS

There has been a strong revival of anti-U.S. sentiment throughout Latin America. This is especially noticeable in the Southern Cone, where former U.S. allies against inroads of communism in the hemisphere have become, under the new "democratic" governments, some of the strongest critics of U.S. policy. How much of this opposition can be attributed to U.S. support of Britain in the Malvinas War, to charges of intervention in Central America, to the Latin American debt crisis, or a combination of all three, is difficult to establish. It does, though, clearly exist, and it seems to be growing under the new democratic regimes.

Moreover, since democratization, "independent," "Third Way," or "nonaligned" policies and actions by democratic governments have become increasingly independent from, critical of, and actively opposed to the U.S. position. In addition to verbal and media attacks, governments are showing their freedom from U.S. influence by taking various policies and actions against the United States. Argentina, Brazil, Uruguay, and Peru participated in the Contadora Group, which sought a Latin American solution to the Central American crisis and the elimination of the U.S. presence there. Brazil has resumed diplomatic relations with Castro's Cuba. Argentina and Brazil are resisting repayment of their debts to the United States, as is Peru, which seems to be rapidly sliding into socialism with the nationalization of its banks and its debt moratorium (an idea it borrowed from Castro and encouraged other countries to follow).

This new independence is illustrated by President Alfonsín's 1986 state-of-the-nation address proclaiming that Argentina belongs to the West culturally, but will not support the West in its military struggle with the East. There is no doubt that this is Argentine policy as evidenced by the growing trade, scientific, and other ties with the Soviet Union. An increasingly independent stance is also evident in Brazil. In 1986 President Sarney visited Washington and was accorded all the courtesies normally given to a visiting head of a friendly state, including meetings with the president and other high officials and the privilege of addressing the Congress. Apparently, this was not enough, as Brazilian newspapers charged that Sarney was not treated with respect by the vice president and the secretary of treasury. Folha de São Paulo declared in an editorial that

> Neither the president nor the Brazilians will bow their heads to gross intimidation by the U.S. government. They will not allow them to submit them to a status of relations that is unworthy of the country's importance, of the country's level of development, and of the country's status as a free and sovereign nation. (Reprinted in the *Times of the Americas*, October 1, 1986.)

While the reasons for the criticism were not given, it is apparent that it was directed at U.S. refusal to accept some facets of President Sarney's proposals for trade and debt agreements. This editorial illustrates a tendency often characteristic of Latin American relations with the United States. A given nation advocates a solution to its own regional or continental problems and demands that the United States accept the proposal verbatim. If the proposal is not approved, it then becomes the basis for additional criticism of the United States, or evidence of the oppression of the South by the North. There is no room left for negotiation, as is the case in many South American disputes. However, these maneuvers do provide the basis for placing the blame for the nation's problems on the United States, a tactic useful in domestic politics.

Fortunately, not all analysts see the United States as the primary threat and clearly identify the causes of national problems and advocate pragmatic solutions. For instance, despite the Alfonsín government's increasing Soviet ties, traditional Argentine scholars continue to emphasize the Soviet threat to Argentina, the South Atlantic transportation links, the Western Hemisphere, and the United States. Villegas (1986a) warns of the Soviet threat, citing USSR incursions into the Third World and the world's oceans as a "war in peace" based on a strategic plan to: (1) maintain the iron curtain in Eastern Europe and attempt to remove the U.S. nuclear potential from Europe; (2) encourage convulsions in the Near and Middle East and Africa to force the West to disperse its forces and efforts in these areas; and (3) control and dominate Black Africa in order to use its ports as bases for the growing Soviet fleet. In this way, Villegas argues:

> The old world, under communism, will encircle the new world and in this geo-ideo-political embrace, the American hemisphere will receive a mortal blow in its most sensitive portion, Latin America. Thus will be consummated the fullest and greatest strategy of envelopment in history. Western Europe will be indefensible, and the United States, surrounded, will remain as an island imprisoned in a Marxist web.

Villegas also charges that Russia is penetrating Latin America with the goal of creating a Marxist integrated and proletariatized continent against the United States. If the Argentine government and people do not recognize the threat and permit time to run without action, their future will be the destruction of their cultural roots and way of life and the loss of their freedom.

In another paper, Villegas (1986b) examines the Soviet menace to the South Atlantic and Antarctic waters, charging that the Soviets seek control of the Atlantic Narrows between Brazil and Africa and of the strategic southern passages linking South Africa and South America to Antarctica. The countries concerned cannot meet the threat by themselves. The leading nations of the western world should ensure the continuity of the independence and the development of the South Atlantic Basin to liberate this region from the Soviet threat and future hegemony.

Villegas represents the older traditional school of nationalist geopolitical thought that tries to arouse both the government and the people against the Soviet threat and to deal with it in cooperation with other western powers. This school recognizes that there can be no security in the Western Hemisphere without the support of the United States. There is another school which emphasizes the contemporary Latin American independence movement (from the United States) and stresses purely Latin American solutions.

The "Third Way," or independent, school of thought is presented by another Argentine analyst, Marini (1986b). After reviewing U.S.–Latin American relations since 1945 up through the period of the National Security Doctrine, Marini argues that the latter only served to preserve the status quo and, with it, the latent Marxist danger, while it failed to solve national social, economic, and development problems or to meet the aspirations of the people. In this climate, international socialism appealed to peoples trying to free themselves from dictatorships. In effect, Latin American nationalist movements, imbued with either a rational or irrational hatred of the United States, reached a tacit understanding and alliance with the Marxists that would lead them toward the Soviet sphere of influence. In attempting to avoid the fall into one or the other camps and the gravitational attraction of bipolarity, Latin American nationalism forgot to create a realistic ideology to improve the future of their peoples. Although the eruption of democracy is a step forward, many so-called democracies are only formal democracies, that is, they have elections, but the people have no say about, or influence on, the government. However, democracy creates an opportunity for a convergence with the United States, provided the latter recognizes the aspirations and desires of the Latin American peoples. The old hegemony-submission relationship and the continuation of dollar diplomacy can only serve to increase Latin hatred toward the United States. The contemporary existence of real Marxist danger alters U.S. relations with Latin America: "For the United States to lose Latin America from the context of its global strategy is suicide." To retain Latin America, Marini contends that the United States must support the Latin American democratic process while these democracies reach their own solutions in their own ways. He urges Latin American integration, but only among democratic states: "It is evident that democratic and dictatorial regimes can not coexist . . . . the integration of Latin America is only possible in democracy." Marini advocates the adoption of the Venezuelan Caraballeda formula, originated at a Central American meeting as the model for Latin American integration and based on these points: (1) a strictly Latin American alternative; (2) self determination of peoples; (3) no intervention in the internal affairs of other countries; (4) territorial integrity in all of the nations of the area; (5) withdrawal from the problems of bipolarity; and (6) democratic pluralism.

Marini believes that, since the United States is now supporting democratic governments, an indirect convergence of the strategic interests of the United States and the Latin American interests of stability and growth can be created. If the United States recognizes the aspirations of and supports the Latin American democracies unreservedly, it can arrive at an effective formula for displacing the Marxists from Latin America. The old hegemony-submission relationship is the exact formula for uniting the Marxists with popular nationalists. Only by supporting national democracies can the United States snatch the banners of liberation from the USSR and remove the only pretext for intervention.

The works of the two Argentine scholars cited above would seem to represent the contending views of Southern Cone nationalists. While both recognize the Marxist-Soviet threat, their frames of reference are very different. Villegas expounds the traditional nationalist position which recognizes the Soviet threat and assumes a convergence of interests between the United States and Latin American nations so as to thwart Soviet goals of Latin American domination in order to envelop the United States. Although he acknowledges U.S. neglect of Latin America, Villegas does not blame the United States for Latin American problems.

Marini's analysis and arguments, on the other hand, represent the ideas and agenda of those who seek a third way, freedom from the problems of bipolarity, that is, from both U.S. and Soviet influence and hegemony. Marini and others charge, that, because of its past support of authoritarian governments, the United States is responsible for creating a community of interest between popular nationalist and Marxist movements. Supporters of this third way acknowledge hatred of the United States and its past and some current policies. The solution they advocate is an integrated Latin America free of both U.S. and Soviet domination. Yet the United States can only achieve a rapprochement by supporting Latin American "democracies" unre-servedly, without considering its own interests. The only U.S. action possible under this concept, then, is that of withdrawing its support for authoritarian regimes. This highly idealized view no doubt represents the

Alfonsín government's position as well as those of some other democratic governments. It appears to minimize the Soviet threat, which presumably will be overcome by "democracy," and to exaggerate the danger of "Yankee imperialism." In this view, the United States must atone for its past sins, but no equivalent demands are made for cessation of Soviet inroads into the hemisphere. The danger here is that opposition to U.S. intervention and "imperialism" will create new opportunities for Soviet penetration of Latin America. The desire for freedom from superpower confrontation is understandable, but this utopian concept is unlikely to come to fruition until an integrated Latin America develops the power and the will to enforce its desired position.

Will the revived anti-U.S. syndrome and the third-way solution reduce or increase conflict in the Southern Cone and throughout the Americas? If a third-way, integrated Latin America could be quickly created, conflict might lessen; if not, as seems probable, the reverse is likely to be true. So far, United States acclamation of Latin American democracies has mostly served to increase Latin criticism and opposition. In this climate, will any action short of a U.S. withdrawal to its own borders and the issuance of a blank check for any and all Latin American initiatives win Latin American approval?

Any solution which ignores the interests, the security, and the well-being of the United States will not serve as a solid basis for a new inter-American understanding. If inter-American unity is not restored and self-proclaimed Latin American independence from U.S. influence intensifies, further Soviet penetration in the hemisphere could lead to more, not less, conflict. To date, based on the available indicators, democratization in Latin America seems to have had a negative effect on U.S.–Latin American relations.

THE IMPACT OF DEMOCRATIZATION

It is evident that a return to democracy has not altered national goals and aspirations. Indeed, in some cases, it may have intensified them, thus preserving old disputes and the seeds of new or renewed conflict. In the case of the Bolivia-Chile-Peru dispute, the last meaningful negotiations took place when all three nations were under authoritarian regimes (from 1975 to 1978). An attempt at talks between Bolivia and Chile in 1987 was quickly broken off without results. The return to democracy in Bolivia and Peru has not changed either government's position, while that of Chile has not changed in the past, regardless of the type of regime in power. An example of "system conflict" was provided by repeated Bolivian attempts to pressure the unpopular Chilean regime in international forums worldwide. This ploy was just as unsuccessful as other Bolivian efforts in the past. To date, the impact of democratization on this dispute is nil.

Disputes of claimed sovereignty over maritime space have not as yet resulted in armed combat, except that ancillary to the Malvinas War, in which the principal conflict was over the islands, not surrounding waters (although, in any case, Argentina failed to control the sea between the mainland and the islands). A clear relationship has been established, however, between claims to sovereignty over maritime areas and other areas of dispute in the Beagle Channel, Malvinas, and Antarctic cases. Here, the possibility of future conflict extends to extracontinental powers, if Southern Cone nations attempt to exercise sovereignty over maritime space. The impact of democratization cannot as yet be evaluated as this case, but the geopolitical concepts involved have been pursued by both authoritarian and democratic regimes since the 1940s and have aroused increasing interest in recent years. Democratic governments show no disposition to withdraw from claims to exclusive national use of maritime space, except perhaps in favor of some regional association designed to exclude noncontinental powers from the region's resources.

The Beagle Channel dispute was settled in 1984, after fourteen years of arguments, threats of war, negotiations, and arbitration. It seems clear that the rise of a democratic government in Argentina had a direct and positive influence on the settlement, although a recently defeated and weakened Argentina may have had no other recourse at the time. Hardline analysts in both countries believe that this dispute is not finally settled; after all, the treaties of 1884, 1893, and 1902 did not represent final solutions. Pertinent to this point is that the 1984 treaty very explicitly recognizes and preserves the geopolitical goals of the two parties, which extend to their Antarctic claims.

In the case of the Malvinas dispute, the election of a democratic government has not changed the Argentine position at all. If anything, the Alfonsín regime has been more vociferous in pressing its claims than have its predecessors. The president proclaims that the Malvinas are, were, and will be Argentine, and he continues to demand that Argentine sovereignty be recognized as a condition for resuming negotiations. Neither the geopolitical goals nor the position of Argentina has changed, and, as long as the present government lacks the military power to force a solution, future conflict cannot be ruled out. The case also illustrates the possibility of conflict involving extracontinental powers, should their interests clash with claims of Southern Cone nations in offshore areas. A democratic government in Argentina seems no more likely to achieve a solution than were its predecessors, for it is following the same course.

The conflicting claims to Antarctica may be the greatest source of future potential conflict because they involve not only Southern Cone national and possible regional claims, but the interests of European and Third World nations and those of the superpowers as well. The Third World, led by such nations as Malaysia, questions whether even the present Antarctic Treaty signatories should be allowed to administer and exploit the area, which they declare to be "the common heritage of mankind" (Luard 1984). Instead, Third World countries advocate creation of a separate regime, directed by a council of perhaps thirty nations, for mineral exploitation and management of Antarctic resources. Their ideas directly oppose the Southern Cone concepts of reserving Antarctic resources for their own or Latin American use. With the Antarctic Treaty eligible for review in 1991, the possibility of continued or future conflict remains open. There is considerable evidence (Child 1985, 1988; Diaz Bessone 1984; Pinochet de la Barra 1984) that Argentina and Chile might welcome an extension of the present treaty, particularly if their claims are preserved. Other measures—such as drastic revision of the treaty regime, or the creation of a separate organization that would include nontreaty signatories in the exploitation of Antarctic resources—would be strongly resisted. While the organization of an Ibero-American Club to represent Latin American interests and claims seems unlikely prior to 1991, such a grouping could be formed at any time in the future if there seemed to be a clear threat to Southern Cone interests. In the case of conflicting claims to Antarctica, the impact of democratization remains to be seen, but it seems implausible that claimant countries will abandon their respective national and collective geopolitical goals and adopt any common heritage-of-mankind solution without a fight, regardless of the type of government in power at the time. As pointed out by Luard (1984, 1173), a conflict of authority in Antarctica could lead to anarchy in the region, with a strong possibility of renewed territorial conflict.

Finally, there is the question of the impact of democratization on U.S.– Latin American relations. The increasingly independent position and often defiant opposition to U.S. policies and actions on the part of the new democracies seems to indicate a worsening of hemisphere relations. The real danger in this situation is that, in their eagerness to achieve a third way solution free of both U.S. and Soviet influence, Latin American democracies may be indulging their anti-U.S. bias to the benefit of the Soviet Union. If this is true, as many believe, the result may be continuing Soviet inroads and penetrations of the Western Hemisphere designed to coopt or dominate Latin America and use it to encircle the United States in its own hemisphere. If this is the case, the possibilities for future conflict are almost limitless, involving Latin American bilateral and regional conflict, inter-American conflict and conflict involving extraregional powers. This situation is complicated by internal dissension and lack of national and hemispheric consensus on the part of both Latin American nations and the United States. Latin American countries, in their search for the best of all possible worlds, may be substituting a perceived, but imaginary, tyranny for a very real one. Can Latin America make good its escape from bipolarity without a struggle?

The evidence, as shown by the ongoing Marxist insurgencies in Central America, Colombia, and Peru, is against such an outcome. Democratic regimes seem to be no better prepared to deal with these problems than were their authoritarian predecessors.

There is also the question of whether the present democratic trend is irreversible. Historically, there have been alternating cycles of democratic and authoritarian rule. A former secretary of state notes that "a reversion to authoritarianism was by no means inconceivable; the political process in the early 1980s was not irreversible (Rogers 1985, 569)." The economic and social problems in the new democracies, insurgencies in Peru, and repeated coup attempts in Argentina, plus growing opposition to President Sarney and his policies in Brazil and the continuing foreign debt problems in all countries indicate that a millennium through democratization has not yet arrived (Graham 1987).

What would be the effect of a return to authoritarian rule in the Southern Cone? Would it make future struggles more likely? It might provoke more internal insurgencies, but it is doubtful that it would increase international conflict. Only the Argentine military threatened war against Chile and engaged in war with Britain over the Malvinas. The Pinochet regime in Chile, although it did mobilize against the Argentine deployment in 1978, has pursued its geopolitical goals through patient negotiations and arbitration, not the use of armed force. The military regime in Brazil used no force against its neighbors. On the other hand, Peru has not changed its geopolitical position under democratic rule, is fighting two internal insurgencies, and, in fact, engaged in combat with Ecuador, another democratic government. The most vociferous and active claimants in current disputes are Bolivia (and its return to the sea) and Argentina (in the Malvinas question), both democracies.

This study indicates that governments in the Southern Cone will pursue their own national interests and geopolitical goals regardless of the type of regime in power and that contemporary democratic governments are often less likely to cooperate with the United States than authoritarian ones. This is why old continental disputes remain active and new ones over sovereignty, the sea, and Antarctica can be expected to arise. To date, even if one credits the return to democracy in Argentina with a settlement in the Beagle Channel case, the impact of democratization on geopolitics and conflict in the Southern Cone has been minimal, and its future impact, by implication, may not lessen tension, conflict, and dispute in the region.

part two

ARGENTINA

chapter three

The South Atlantic: War Hypothesis

Norberto Ceresole

THE SOUTH ATLANTIC IN THE EAST-WEST CONFRONTATION

The 1982 South Atlantic war stimulated in the heart of the North Atlantic Treaty Organization (NATO) a debate that up to that time had been dead, or at best lethargic: the debate dealt with a reassessment of the strategic value of the South Atlantic.

Even toward the end of the 1970s the South Atlantic was generally considered to be a great strategic vacuum, "a vacuum of power, of domination, and of control" (Salgado Alba 1984). But even then there were warnings: "The existence of a vacuum always suggests a danger, a risk, or at least a breeding ground of tensions." Besides being a "power vacuum," the South Atlantic was also "an intellectual vacuum of Western geo-strategic thinking." All of this gave rise to the question: "Who will fill the vacuum of power, of control, and of domination that is the South Atlantic today?" (Salgado Alba 1984).

From 1982 on, it has been possible to note a change in NATO's perception of this regional problem. Slowly but surely, this enormous maritime zone and its coasts and adjacent territories are acquiring a new dimension in strategic thinking, from the Atlanticist perspective. The problems of the South Atlantic are beginning to be more clearly perceived as having a relationship to security problems of the North Atlantic (Rosch 1984).

This Atlanticist policy is based on the equivalence of both North and South Atlantic security modules (and this, in turn, assumes that one accepts *a priori* the hegemony of North Atlantic interests over South Atlantic ones). The security threat stems specifically from the Soviet fleet (surface and

Translated by Jack Child

submarine) and is basically related to the following strategic equation: whoever dominates South Atlantic traffic will dominate northern maritime Europe and the "U.S. Atlantic."

Argentine geopolitical thinking has traditionally accepted the concept of a South Atlantic from that same "Atlanticist" perspective. That is to say, it viewed the South Atlantic as a maritime containment barrier against Soviet "expansionism" coming out of the Indian Ocean basin, West Africa, and the Southern Pacific Ocean (Indochina Peninsula). All that Argentine civilian and military "strategists" asked for (and continue to ask for) was a political arrangement with Great Britain for the recovery of formal sovereignty of the Malvinas Islands. This would have presupposed an automatic commitment on the part of Argentina to jointly transform the Islands into a military base against "the growing Soviet naval presence in all the oceans of the world" (Mastrorilli 1977).

Basically, the proposal was that priority should be given to "certain points that might have been overlooked by European observers, who were mainly influenced by the rapid succession of African developments, whose principal impact consisted of the danger of sovietizing vast coastal sectors of southwest Africa." The reinforcement of the South Atlantic's American coast would have as its objective to stop the "red tide" which was taking over vast African regions. For this reason, Argentina was supposed to (and would still) support NATO interests and the "strategic integrity of NATO." Clearly, this would follow the joint construction of "a great naval base on the Malvinas Islands." But this project was blocked by the presence of Great Britain on the archipelago (Mastrorilli 1977).

Although it may be hard to believe, this perception of the South Atlantic persists at the strategic level in Argentina even after the humiliation, the fracture, and the territorial theft that the West has wrought upon the country. Even after the 1982 defeat, this conception was especially visible in the Navy, which, pathologically, insisted on conceiving of Argentina as an "integral maritime country." As a result, they felt that the international role of the country should be based on its being a part of the "free world in order to act according to the spirit of the West" (Lanzarini 1982). And these principles would be developed along specific courses of action. The role of Argentina in the South Atlantic would be structured as a function of the containment policy which the capitalist maritime world would generate with regard to the continental world of the Soviet Union. This would be especially true because: "the importance of the South Atlantic to Argentina is based on the fact that first, from its bases in Africa the Soviet Union can operate on our territory or that of our neighbors, and second, through the South Atlantic flows maritime traffic which is vital to the Free World. . ." (Lanzarini 1982).

There obviously exist direct and close relationships between and among the counterinsurgency actions of our military institutions, their operational impotence, and their lack of a will to fight an external enemy that operates under the protection of an ideological legitimacy (anticommunism). This ideology is shared by hegemonic sectors of Argentine society and is used as a replacement for national strategic thinking. Ideology replaces strategy, and the nation loses its direction.

After 1982 the Falkland Islands have been welded to the global Atlanticist containment strategy, which is articulated on the periphery of the continental world and on the maritime (aero-naval) cracks which it causes in that containment strategy.

This welding has as its basic objective the filling of a strategic power vacuum. From this point on, the South Atlantic region, in effect, becomes part of changing global politics. That is to say, it becomes an organic part of the principal contradiction of the contemporary world, which is that determined by the so-called East-West conflict.

If the Argentine nation hopes to survive and become a viable country, it must begin to think of its national role within the framework of the global conflict that it is now a part of. The issue is no longer an ideological problem that might serve to destroy an enemy located inside Argentine society.

The South Atlantic is within the cosmovision and the hegemonic needs of the maritime world. It is a maritime zone, but it also includes a large mass of Argentine continental space (Patagonia), and it has been transformed into a shatterbelt:

> The geostrategic vision based on the theoretical assumptions of three great figures of classical geopolitics, Mahan, Mackinder and Spykman, has been synthesized by Bernard Cohen, who now inspires U.S. strategic thought. The vision divides our planet in three great zones, or geopolitical "worlds". These are the Maritime World, with its center in the United States, and which depends on commerce; the Euroasiatic Continental World, with its center in the Soviet Union; and the Indian Ocean Basin World, presently being formed. Between these 'geostrategic worlds', according to Cohen, there exist the so-called 'shatter-belts', which are zones of tension and instability that separate them (Salgado Alba 1984).

A shatterbelt in the interior of our national space is the clearest possible threat to our real interests in that space and is the clearest and most precise factor which threatens our security.

The existence of an interior shatterbelt in effect fragments Argentine space and eliminates any possibility of exercising sovereignty. This shatterbelt placed in the interior of our space is the result of: (1) a native armed force conceived as a function of ideology and oriented toward the destruction of interior enemies; (2) the global military policy of the capitalist maritime world; and (3) the subordinate status which that world assigns Argentine space, relegating it and placing it, literally, on its periphery (Cohen 1980).

This global military policy of the West for the region has a basic requirement: the space under consideration must be susceptible to complete domination. Spaces are conceived as the organization of "a single zone of movement . . . (and) the unity of the zone of movement is indivisible." The doctrine cannot conceive of any type of discontinuity in the exercise of power over the surface of that space and must attempt to minimize any such discontinuity. Any possibility of sharing the advantages offered by those areas with "secondary users" is perceived as a fracture of the zone, understood as a unit of movement (Cohen 1980).

If the Malvinas Islands are seen as part of an "impact zone" within the South Atlantic (and this is our hypothesis), then not only is the exercise of Argentine sovereignty over them made impossible, but, further, any agreement with third parties would be perceived as an aggression. The limiting point of this reasoning is reached when the third party is the superpower of the territorial heartland (Schultz 1986). If this reasoning is correct, then the strategic error committed by the Argentine government has reached colossal dimensions, from the moment when its foreign policy has become centered on the attempt to be seen as reliable within the Western world.

If Argentine sovereignty over a portion of that space is perceived as a fracture of the uniform module for the exercise of power, then there can be no room, within the West, for the exercise of Argentine sovereignty over its own territory. This is because that exercise of sovereignty is perceived as a fracture in the zone of movement and as a reduction of the strategic value of a specific space in a global conflict. This real situation, in its concrete expression, has a logic which must be followed to its final consequences. The destiny of the nation is linked to it.

Such an impossibility of having sovereignty over one's own space (an impossibility caused by our alignment within the capitalist maritime world) creates an unbearable threat not only for the security, but also for the very survival of the nation. No nation can simply exist with a shatterbelt located in the interior of its own space.

The survival of Argentina will therefore depend on the formulation of a war hypothesis aimed at transforming that shatterbelt into a space under our effective sovereignty. This assessment, with identical logical validity, can be expressed inversely: from the point of view of Argentine national spatial interests, any fracture caused by third parties in the zone under consideration will automatically have benefits for us. Under present conditions, given the military hegemony which Western powers exercise over the region, any fracture which is introduced and which would be a discontinuity in the exercise of power by the capitalist maritime world, is not only beneficial for our national interests, but is also a basic condition for our own power that can mean, at some point, the sovereign use of that strategic space.

There is a direct and immediate relationship between, on the one hand,

Argentina's national capability to exercise sovereignty and, on the other hand, the fracture of the South Atlantic's maritime space (in relation to NATO's unity of power and movement) that the Soviet Union's political and military power might produce. From this stems a conclusion: the elements of fracture and discontinuity the Soviet naval presence in the region might introduce, will be, in all cases, objectively favorable to Argentine national interests, regardless of the ideological perception one might hold of them.

Nevertheless, these "external fractures," which can occur in the zone under its present hegemonic configuration, do not automatically solve the issue of Argentine sovereignty over that zone. They can only create favorable conditions for a political decision, for an act of national will that still does not exist, which would carry out and give shape to a historical project. External conditions, regardless of what they might be, cannot replace national policies, unless one simply wants to change from one "ideological world" to another. And that is not the issue here.

We are speaking of a maritime theater of operations (which might in some cases also be aero-naval), and therefore we must make reference to instruments and forms of action represented by fleets: air, as well as naval surface and submarine ones. But the zone is not simply a naval/air theater of operations; it is also a continental one. The shatterbelt not only fractures maritime and air space (in its present military and strategic configuration), but it also denies the use of that space to a decisionmaking system centered in Argentina. Further, it makes unviable enormous areas of our country that are strictly continental, beginning with our Patagonian space. It blocks the nation from making full use of our own space because it is subjected to a threat that degrades the situational value of Argentine continental-territorial space.

This relation between external threat and degradation of strategic value, in that specific relationship, had been perceived with full clarity by Admiral Storni, who wrote in 1916 (Storni 1952):

> The building of a powerful foreign naval base on the Malvinas Islands could not help having, as one of its immediate consequences, the domination of our adjacent seas *and their coasts.* . . If the Islands remain in foreign hands for an indefinite time, then we will not be able to fully resolve the problem of our maritime defense, *no matter how much we perfect the defense of our mainland coasts* [our italics].

This projection of the potential power of "Fortress Falklands" toward Argentine continental space (in the first instance toward Patagonia) has no relationship to the power of an "Argentine threat" against the islands (assuming that today we indeed have national military power). The growth of the military capability of the Falklands is a factor directly proportional to a global power relationship. That military capability depends, both qualitatively and quantitatively, on the East-West confrontation.

That is why Argentina is already fully participating in one of the fronts of global confrontation. Today in the South Atlantic one can see, in a specific form determined by the nature of the theater of operations, a modality of confrontation that up to now had been relatively distant from Argentine concerns. What had been only an ideological problem administered by the anticommunism of the local managerial classes, has been suddenly transformed into a strategic and military problem of growing magnitude. The projection of the center of gravity of world problems is there, just around the corner.

Argentina must now definitely make a choice. It will no longer be possible to remain on the margins of East-West conflict. The conflict is now within our territory, and it affects our national viability. In this conflict one side is objectively an aggressor of our country, and a system of forces that is the strategic opponent of our aggressor is our natural ally.

There is only one way to exploit to the maximum the new system of alliances with regard to the imposition of geopolitical surrender of our national territory caused by the military aggression of the West. This path is the one in which Argentina becomes increasingly autonomous and increasingly realizes its full potential (Ceresole 1984). This would be an Argentina growing stronger internally and clearly breaking away from the traditional forms of its foreign policy. Only in this manner could we establish the basis for a successful exploitation of the new international circumstances brought about by Western military aggression in the South Atlantic.

Achieving the full potential of our national space means changing *all* current trends at *all* levels of activity in the country. From the enormous range of factors involved in increasing the potential of our national space, in this present paper we will mention only those relating to defense. We begin by stressing that the democratic system, under the present strategic conditions, cannot survive if it does not assume a specific dimension of power. Without an autonomous and controlled national space (with indivisible sovereignty over power and movement), no policy is possible, because the survival of nations is fundamentally linked to power politics. The international system is an interaction of power politics: the only response to power is power.

A new defense policy, based on our own industrial and technological capabilities and on a profound reform of military institutions, must have as a logical corollary a foreign policy that is absolutely unrestricted and de-ideologized. Furthermore, it must exploit to the fullest the new strategic conditions the nation has been experiencing ever since the imposition of the current aggression, and it must develop a system of South-South alliances. This dimension of our foreign policy, with regard to South America, must lead to the creation of an independent geopolitical region (ILCTRI 1983) in a space which, given certain historical and geographic circumstances, has shown itself to be a unit of movement. This region would basically be made up of six states (Uruguay, Paraguay, Bolivia, Peru, Chile, and Argentina) and would have all the constituent elements of an independent geopolitical region: a principal ecumene, empty spaces, access to two oceans, and a high degree of continentality.

An independent geopolitical region so defined would imply perceiving the South Atlantic as an element of linkage and continuity between two ocean basins that would acquire a high strategic value for Argentina: the Pacific and the Indian. And in both directions the independent geopolitical region would find that the Soviet ocean fleet would no longer be the "opponent," but rather a channel of movement (Cohen 1980). As such, it would increase the natural potential of the region's autonomous capabilities and its movement at the strategic level. The independent geopolitical region's own maritime capability, designed as a function of its real economic needs, could not be thought of as having strategic interests opposed to the maritime capability of the continental world. In this manner, the old intuition of Mackinder (1975) would be fulfilled:

> The development of the great potential of South America could have a decisive influence in the system . . . . my point is that from a geographic perspective those great potentials probably will have to revolve around the Pivot State (Continental World--Territorial Heartland), which will probably always have to be large, but with limited mobility if it is compared to that of the marginal and insular powers which surround it.

For Mackinder, this would introduce an important alteration in the international system, since it adds an "oceanic front to the resources of the great continent" (Mackinder 1975). Current sovereignty problems in the South Atlantic are not circumstantial political problems. They are strategic problems that will have an impact in the very long range.

The objective insertion of Argentina into the nucleus of the East-West conflict subordinates all earlier conflict hypotheses to the present one, which is projected to a logical higher level: a conflict hypothesis leading to a war hypothesis. In the South Atlantic theater of operations we can project a global confrontation between the maritime world and the continental world.

The traditional conflict hypothesis which Argentina maintained with regard to Brazil (and vice versa) assumes at present a different dimension. Brazil's new policy in the South Atlantic is oriented toward developing a "zone of peace." This implies the "demilitarization" of the region, that is to say, the creation of a "military vacuum" in that vast surface of the planet. For obvious reasons, no region can remain militarily empty in today's world. Therefore, theoretically, its military "filling" would be the responsibility of the coastal states with direct and immediate interests in the region (east coast of America and west coast of Africa). But in real terms, and given the foreseeable "loss" of South Africa, the only regional power with effective capability to militarily fill the region is Brazil, especially if one considers

Argentina's current "defense policy," and its obsessive de-industrialization (Ceresole 1986).

Since the end of the 1982 war, Argentina has been moving, from the defense viewpoint, along a two-way street. Together, these two directions have been catastrophic for the true security of the nation. On the one hand, there is an increase in the defenselessness of the nation with regard to the external enemy (the projection of power by the West over the region). On the other hand, the archaic military structures have not been dismantled, and these continue to act violently against a presumed "internal enemy" with "communist" roots. As a result, Argentina has the worst of all possible worlds.

Argentine military defenselessness with respect to her external enemies, which can be readily characterized and measured, is the result of various factors: (1) a falsely pacifist "world view" totally assumed by the constitutional government; (2) a doctrine of "universal democraticism" ("all democratic states are our natural allies"), which attempts to occupy the space left vacant by the departed strategic concept of the "national security doctrine"; (3) an obsession with de-industrializing the defense sector (and this, among other things, carries with it the passive acceptance of a regional division of labor in this field); (4) a lack of interest in military policy as such, in its organizational and institutional aspects.

For all of these reasons, Argentina faces a new phase in the South Atlantic conflict, without a defense policy, without military doctrine, lacking the appropriate weapons, without an economic or foreign policy that would protect it from the aggressive interests of the West, without the appropriate institutional instruments (military organization), and with armed institutions solidly tied to the most aberrant forms of counterinsurgency doctrine. With this heavy load of baggage, the government proposes "global" negotiations regarding the region at the lowest imaginable level: accepting the existence of Fortress Falklands and multilateralizing the diplomatic dispute in accordance with the British conception.

One cannot doubt that the greatest interest of present Argentine foreign policy is focused on "confidence-building" within the West. This will bring very serious consequences both internally and externally. At the level of internal policy the political initiative will be consolidated in the hands of the established system. At the external level, there will be an increase of Argentina's alignment with the "guiding powers of the West." There is a possibility that, in the case of the regional conflict we are dealing with here, the Argentine democratic government will become a junior partner in the Atlanticist project in the South Atlantic (after the return of some formal symbols). In the midterm, Argentina, on behalf of the West, may also become the base from which to deal with a low-intensity conflict in Chile.

A FUTURE SCENARIO

The ideas laid out in the prior section demand the design of an alternate "future scenario." The present scenario is understood to be the basis for a foreign policy that increases the nation's weakness in defense matters and that, in the broadest sense, serves to drive Argentina's decline.

We will attempt to design this future scenario as the external and objective basis for an unrestricted and unaligned Argentine foreign policy. Such a scenario could be structured in the following manner:

1. The basic contradiction at the global level is the East-West conflict, from the moment this conflict can provide or deny any alternate policy covering the whole planet, and even beyond.

2. There is no North-South conflict, but rather a South-Northwest split. Underdevelopment in any of its manifestations is the other side (and the very condition of the existence) of industrial and democratic capitalism (of the capitalist maritime world). South-Northeast conflicts are qualitatively different. The Northeast has no need to condemn the South to underdevelopment. The logic of the system is of a different nature.

3. In assuming that the South-Northwest conflict is distinctively Argentina's, we do not imply that the country is outside of the principal conflict (the East-West conflict). Rather, we imply that both systems of contradictions impact on Argentine national reality. In any event, Argentina must process both conflicts simultaneously and, from that point, assume a new perspective on the situation.

4. The "North" is a "world" culturally as well as geographically and politically divided. Geographic discontinuities separate the capitalist maritime world from the socialist continental world. These discontinuities establish fracture lines and differentiations that very clearly form two basic areas of movement.

5. The capitalist maritime world (CMW) is structured into hierarchically organized geopolitical subareas or regions. From the viewpoint of the center of the system, our country is located in an absolutely marginal zone, in the region of least strategic interest. From a geographic and cultural perspective, Argentina is, strictly speaking, the periphery of the system.

6. The different hierarchical levels express different degrees of availability of power. This is the root of South-Northwest conflicts. For our position, on the periphery of the CMW system, it is fundamental to perceive (and to be able to act on accordingly) the nature of the phenomena that gives rise to those levels of hierarchy (levels of power) within the system. The effective rupturing of this hierarchical scale of the spaces with less available power, located in zones of "low strategic interest," cannot be a process within the system. Thus, the resolution of the South-Northwest conflict must basically involve the resolution of the East-West conflict. This is especially

true for those countries, such as Argentina, that are geopolitically far from the center.

7. The movement from a position of geographic marginality (with extremely low power profiles, which is "fatally" dependent and marginal) to a different geopolitical position, with some capability for autonomous action, is not an unsolvable problem framed by an absolutely anachronistic geographic determinism. If one begins by absorbing technologies allowing policentric possibilities (Ceresole 1983), it is possible to develop a capacity for politico-strategic action. This requires a perception that it is necessary to give our space a situational value. This value is absolutely different from that which emerges from the geographic "determinism" pushing us to the world's periphery and assigning us infinitesimal quotas of power.

8. The break with the determinism of our geographic "sentence" is in the first place an act of strategic will that must be deeply rooted in the heart of our people. Only by starting from this point will it be possible to develop two parallel courses of action over time. One of these courses of action works within the regional political environment and deals with giving power and structure to this regional space so as to shape it into a single unit of movement within the geopolitical scale. The second course of action, inseparable from the first, is to manage to link that regional space (which we have called an independent geopolitical region) to those most dynamic elements in the geostrategic zone opposed to the CMW (that is to say, with the socialist continental world) using the maritime (bi-oceanic) condition that would be achieved by the independent geopolitical region, which we have proposed as a unit of movement.

9. The two courses of action mentioned above—one at the regional level, and the second at the global—are imbedded in two different geographic orders or modules. The first is found at the South-South level (which, of course, could be broadened to exchange technology with countries like India). The second is located at the South-Northeast level (South-Continental World, as Mackinder intuited). It is not possible to carry out both movements separately, or at different times. In turn, both together will make possible other maneuvers which are absolutely essential for the Latin American Southern Cone's independent geopolitical region, such as closer ties to the Indian Basin.

10. The second course of action mentioned in (8) has an eminently geostrategic dimension. It consists of connecting the independent geopolitical region with dynamic elements of the continental world. This linkage is possible only if it is based on the high potential for mobility that access to both oceans gives the independent geopolitical region. This mobility turns out to be absolutely consistent with the existence of a Soviet ocean fleet capable of rupturing the maritime world as an area of unity of movement. Such a course of action not only does not exclude the previous one, but also is an essential condition in order to carry out movements of international

politics in the "horizontal" South-South belt. The reality of East-West confrontation and the different nature of South- Northwest and South-Northeast relations make possible the existence of power vacuums at the surface level of world politics. These power vacuums can give rise to revolutionary processes and, as a result, to the emergence of new international forces. Seen from the interior of the CMW, the East-West confrontation is what makes the power of the center somewhat different from its capacity to issue orders and ensure that they are obeyed on the periphery. The power/ order dichotomy is born and develops—to the benefit of the emerging peoples—thanks to the characteristics taken on by the East-West conflict.

A different perception of the situational value of our space—as we have attempted to point out previously—brings with it the design of an Argentine foreign policy based on assumptions partially or totally different from those that govern the present reinsertion of Argentina into the West. The reevaluation of our national space could then coincide with:

1. A growing decline of the U.S. economy: This would bring with it, in parallel, a greater strategic weakness in the ecumene of the CMW, independent of the adoption of new military technologies. This decline leads to a growing vulnerability of the Western economic system, to its internal instability, to a lack of leadership and adequate hegemony within the inter-capitalist competition, and to the lack of adequate answers to the new features of the international economic crisis. All of this, among other things, promotes growing heterogeneity in the forms and rhythms of growth. The global and regional repercussions of this situation are multiple. They include the emergence of new forms of warfare, such as the so-called low-intensity conflicts.

2. The hypothesis indicated above would lead to an important redistribution of relative power within the CMW. A decline in U.S. power would be accompanied by a greater competition among the capitalist states, which would include not only the highly developed states such as Germany and Japan, but also emerging powers such as Brazil, some NICs, and so on.

3. In a similar fashion, it is possible to foresee the consolidation of certain economic, technological, and military developments in the socialist continental world (SCW), such as the blossoming of outer space technology and the quantitative and qualitative increase in the movement capability of the soviet naval fleet.

4. It is very likely that there will be profound geopolitical changes in the Pacific Basin. A series of recent indicators suggest the probability of general destabilization (Institut du Pacifique 1983). This would fracture certain Asian offshore zones created as containment barriers by the maritime world in order to surround the continental heartland. Here the key point is determined by the situation in the Philippines. Their fall would give the soviet fleet a mobility which has been absolutely unknown to this moment.

5. The evolution of the internal situation in Afghanistan would make it possible to foresee the physical and geographic welding of the continental heartland and the monsoon region of Asia. This would suggest the hypothesis of a growing deterioration of the internal situation in Pakistan and increasing instability in the Gulf states, which would give even greater value to the Indian/Atlantic interoceanic passage.

6. The relative decline in the economic and general situation of the United States would give new value to the location of the Central European states. These would once again act as links between the two Europes, as an intercontinental bridge. In Central Europe a geopolitical situation could reemerge similar to that of the period after World War I, with the border of maritime Europe located on the eastern frontier of France. Within this context, nations such as Poland and the two Germanies would be main actors in a movement of European recontinentalization. The "Yalta frontiers" would have demonstrated their enormous political utility and their total historical validity, at least in the region to which we refer. It was Yalta, for example, that gave the Polish state something it never (or almost never) had in two thousand years: territorial security and well-defined frontiers—in sum, national existence.

7. Within the scenario we are describing, more favorable conditions would be generated for the achievement of a bipolar balance in the interior of the continental world between the Soviet Union and China. This is one of the basic elements of the international situation at the strategic level. A relative decline of U.S. power would consolidate centripetal forces in the continental world.

This last point would stimulate new conditions in the Pacific Basin, which could be transformed into the principal movement area of the independent geopolitical region of the historical Latin American Southern Cone. As a result, any analytical treatment dealing with the South Atlantic could never be decoupled from that of the South Pacific. In a very strict sense, the new Argentine geopolitics proposed from this institute (ILCTRI) will consider issues in both basins as well as their natural extension, the Indian Ocean.

Thus, from a "marginal" geographic perspective, the increase in the Soviet ocean fleet's mobility would represent a "movement channel" which would increase the maritime power of the independent geopolitical region.

chapter four

Argentina: Ten Years of Foreign Policy Toward the Southern Cone

Roberto Russell

This chapter deals with the foreign policy of Argentina toward the Latin American Southern Cone and Brazil from the beginning of the so-called Process of National Reorganization in March of 1976 until the end of 1987. It thus includes the first four years of the Alfonsín administration.

Given the space limitations, this chapter will not present a minutely detailed analysis of that policy. Instead, it will concentrate on the various assumptions that sustained the policies of the military autocracy and the democratic government that followed, toward the countries of the region. It will also present an analysis of certain special cases useful in showing the profound differences that exist between the two regimes.

In order to achieve a greater approximation of the military government's foreign policy toward the Latin American Southern Cone, it will be necessary to introduce a brief summary of some of the internal variables which affected the overall foreign policy of the regime, which was not exactly characterized by consistency, continuity, and predictability.

When analyzing the regime's decisionmaking process it is necessary to consider the intramilitary conflicts that were a constant in the Process of National Reorganization. These conflicts were the result of both personal struggles to control greater power (for example, those produced between Videla and Massera or between Galtieri and Viola) and differences regarding the government's own actions, especially those involving the economic model promoted by Videla and Martinez de Hoz.

It is also necessary to bear in mind that the success of the economic plan designed during the consolidation period of the authoritarian military regime required the maintenance of a close alliance between the armed forces (in charge of repression) and a new hegemonic coalition (exporters, importers,

---

Translated by Jack Child

and the upper strata of the national and international bourgeois, led by the financial sector). This coalition was prepared to definitively push aside another coalition, which had grown in the shadow of populism, and was made up of national entrepreneurial sectors, middle class groups, and organized labor.[1] The new alliance, based on the economic model designed by Martinez de Hoz, nevertheless contained profound incompatibilities with the long-range geostrategic and national security objectives of the armed forces. In effect, the ideology of the "magic of the marketplace," of free exchange, of comparative advantage, and of the "shrinking of the State in order to make the Nation greater" (which was perhaps the slogan most used by the Process), would turn out to be difficult to harmonize with the "nationalist" ideology of important sectors of the armed forces.[2] An understanding of the "structural tension" existing between the divergent objectives of the central components of the alliance between the armed forces and the new hegemonic coalition is essential to explain, to a large degree, the dynamics and contradictions of the regime in both the internal and international spheres.

Despite the obvious nature of this tension, it was not perceived by the majority of the Argentine armed forces, at least in the first stage of the Process. As Tulchin indicates, it appears evident that the

> Junta never contemplated the contradictions between the economic and strategic long-range requirements of a national security regime, and the painful process of re-inserting the Argentine economy into the international market on the basis of an assumed comparative advantage in the production of certain agricultural raw materials (Tulchin 1984, 370).

As the months passed, this situation began to consolidate and define various conflicts. In the specific plane of foreign policy, it shaped the formation of two diplomacies: one economic, the other military. Both had diverse orientations; both were implemented by different actors and attempted, in general terms, to achieve objectives in great measure incompatible.[3]

In broad terms, the principal features of economic diplomacy were:

(a) A pragmatic orientation, which was internationalist and "First World"

(b) The existence of close ties with the value system and interests of the "world of business" of the developed capitalist countries

(c) Acceptance of the international status quo, whose rules of play Argentina could not and should not substantially change

(d) The support of the principle of the subordination of the State, and the defense to the bitter end of the "magic of the marketplace"

(e) Favoring of a greater internationalization of the productive structure and a consequent perception of the territorial boundaries of the national state as obstacles that limit the "healthy" development of the laws of the marketplace at the global level

(f) Faith in the profit motive as the essential moving force of human actions: private vice as a public virtue which definitively tends to maximize the general welfare

For its part, military diplomacy was characterized by:

(a) An ideological orientation, which was nationalistic, and "Western"
(b) Support of power politics, in many cases in its most extreme forms
(c) Acceptance in general terms of the global status quo, even though the expansionist policies of international communism were severely condemned
(d) Predominance of "activist" nationalist currents that proposed for Argentina (given the country's objective conditions) a path to its "historical destiny" in Latin America
(e) The need to definitively consolidate the integration of the nation's territory
(f) The assignment of an important role to the state in economic development, which, in accordance with the teachings of the national security doctrine, was considered an essential process in avoiding the outbreak of new terrorist groups
(g) The theoretical rejection of the profit motive as the basic motivation for human actions

Military diplomacy assumed control of fundamental foreign policy matters during the whole period of the Process: the southern conflict with Chile, the Malvinas question, the topic of shared natural resources with Brazil, interventions in Bolivia and Central America, and matters related to political and strategic ties with the United States. "Low politics" was left to economic diplomacy, which relied on considerable power and excellent international links and which, as a result, was carried out almost independent of the military authorities.

Finally, it is necessary to note that in this period the armed forces acquired effective autonomy from the dominant economic groups, managing to significantly increase their corporate privileges and impose, to a great extent, their own perceptions and visions of the world. As a result, it would be excessively simplistic to assume that the military were merely agents of the exclusive interests of a given social class. This point seems relevant in order to increase our understanding of the confrontations produced within the regime and the erratic course of its foreign policy.

In spite of its notorious inconsistencies, the military government articulated its international actions on the basis of four principal elements: strong adherence to the theoretical assumptions of the East-West model; fundamentally "mercantilistic" links to Latin America, accompanied in some cases by an expansionistic territorialist nationalism and by a novel interventionism aimed at blocking the advance of "Marxist internationalist" subversion in the region; a low profile in the fora and organizations dominated by the developed nations; and clear pragmatism in commercial and economic matters.

The first three of these features were completely set aside with the coming of democracy. In its place, the foreign policy plans of the Alfonsín government have centered on the following guidelines:

(a) Cultural adherence to the West and opposition to the merely strategic components of the East-West conflict

(b) Reconsideration of Argentina's active participation in the Non-Aligned Movement based on a return to the founding principles of the movement

(c) Support for various schemes for selective cooperation and integration with countries of the developing world, especially Latin America, within the framework of a program for greater autonomy.[4]

THE NATIONAL REORGANIZATION PROCESS

After the 1976 coup, the Process returned to traditional low-profile approaches toward Latin America. To be sure, this policy was strictly consistent with the rigid geopolitical visions of important military sectors, as well as with the interest systems of the principal actors of economic diplomacy, who hold a structural suspicion toward regional collective and concentrated mechanisms.

Nevertheless, as has been stated, this policy was affected by two components contributed by military diplomacy: the "Western" interventionism destined to maintain individual and collective security in the face of the "international communist threat"[5] and the "territorialist" nationalism whose priorities in the external sphere were to secure the borders of the nation (especially in the Beagle situation), to militarily recover the islands of the South Atlantic that were in British hands and, finally, to occupy the "empty spaces" left behind in the wake of the "declining imperial power of the United States."

Each one of these components, singly or in combination, directed a substantial part of the regime's external actions toward Latin America. Thus, in practice, the struggle against "international subversion" led important sectors of military diplomacy to ignore national frontiers and the principles of nonintervention and self-determination. For its part, "territorialist" nationalism produced one war, almost unleashed another, and, combined with "Western" interventionism, led to the unfortunate intervention in Bolivia in July of 1980.

The policy of securing national frontiers found its clearest manifestation in relations with Chile as a result of the so-called Austral (southern) question. Although this problem has roots that go back to the end of the last century, the most significant events were precipitated by the signing in London (on July 22, 1971) of an arbitration commitment between Argentina and Chile. The border question would be submitted for consideration by an arbitration court made up of five judges of the International Court of Justice, whose verdict would be sent to the British Crown to be accepted or rejected as a whole.

The arbitration award was accepted by Chile and rejected as irreversibly null and void by Argentina on January 25, 1978 on the grounds (among others) that it contained "gross errors," "violations of essential juridical rules," and that it "exceeded its authority," since the award arrived at conclusions affecting geographic regions beyond the area it was supposed to consider. Tensions between the two countries increased and reached their highest point toward the end of 1978.[6] In order to avoid an apparently imminent war whose need and "rightness" had been fomented by the nation's military leadership, both parties agreed, by means of the Acts of Montevideo (January 8, 1979), to submit the matter to papal mediation, which John Paul II accepted on the 25th of the same month.

The Montevideo agreement was important because both countries submitted to Papal arbitration the totality of the Austral zone, which covers an area much broader than the "hammer" (a hammer-shaped polygon including all of the Beagle Channel and the contiguous ocean triangle). Thus, they committed themselves to seeking a total and definitive delimitation of the whole disputed area, both maritime and land. In the words of Videla, then president of Argentina: "We have spoken of the Austral zone, and thus we are not limited just to the 'hammer.'"[7]

Within this framework, on December 12, 1980, the pontiff delivered his peace proposal to the diplomatic representatives of Argentina and Chile. Unfortunately, this proposal was not able to break the deadlocked situation.[8] As a result, the matter remained in hibernation in the Argentine agenda toward Latin America until the arrival of the democratic government.

In the Process of National Reorganization, the second key aspect of foreign policy toward the Southern Cone was the participation of military diplomacy in the García Meza coup in Bolivia in July 1980. As mentioned previously, the intervention in Bolivia is a prime example of the combination of two components: "Western" interventionism and "territorialist" nationalism.

With regard to the first component, the Bolivian adventure expressed, better than in any other case, the search for "national" objectives considered vital in order to strengthen the internal control system and block the development of new "terrorist outbreaks."[9] It is worth emphasizing that the objectives involved in the Bolivian intervention did not at that moment coincide with those of the U.S. administration. In effect, the electoral victory of Hernán Siles Zuazo's Popular Democratic Union (UDP, from its Spanish initials for Unión Democrática Popular), which included the most important groups of the Bolivian left (Movimiento Nacionalista Revolucionario de Izquierda, Movimiento de Izquierda Revolucionario, Partido Socialista Uno, and Partido Comunista Boliviano), was perceived by the Argentine government as a threat to regional security and as a sort of Trojan horse that would institutionalize "international subversion." In the perspective of the Argentine and Bolivian military, there was a need to continue the struggle against that subversion (which had been almost wiped out in Argentina at the military level). The "danger" of the electoral victory of the Bolivian left brought together the military establishments of the two countries in their struggle against the common enemy of "international Marxism" and its internal allies. Videla stated in August 1980 that

> What really happened in Bolivia was a choice between two options in that neighboring country. The first choice was the formally correct one of a government coming to power through elections, but which represented for us a high degree of risk because of the possibility of the spreading of ideas contrary to our way of life. The second option was a military government. We favored the latter option because we do not want to have in South America a government that would be able to act like Cuba does in Central America. . . . We are not simply helping the Bolivian military, we are helping the Bolivian people so that they will not fall into something that we almost fell into.[10]

At the same time, the Bolivian operation was conceived as part of a growing external action, whose ultimate destiny was the filling of "power vacuums." In this regard, it is interesting to note that toward the end of the 1970s some of the plans prepared by the general staffs of the Argentine armed forces foresaw (within the framework of the most refined theory of "realism") the unavoidable reduction of the relative power of the United States would force that country to reduce its interests and commitments in the Southern Cone of Latin America, which was a "non-vital" sector of its empire. In this context, the involvement in Bolivia sketched the outlines of a reality that, two years later (in 1982), would end up persuading even the most incredulous: the plans of the military junta, far from being those of simple men on horseback, that include as a fundamental parameter an expansionist nationalism which would lead to a reordering of Argentina's position in the world and in the Latin American environment (Domínguez Ruiz 1983, 97–98).

As a result, the Bolivian question became transformed into one of the key issues of military diplomacy. Its undoubtable importance became evident through unconditional support of the de facto government and an active participation in Bolivian political events. This participation was carried out at various levels: organization of antiguerrilla groups, the sending of personnel specialized in "antisubversive" tasks, economic aid, and the diplomatic and political protection of the García Meza regime. Argentina was the first country in the world to recognize the military government of Bolivia.

Nevertheless, the growing resistance of the forces of political opposition in Bolivia, and the later deviations of the regime, created obstacles for the bilateral "military understanding." The deviations of the Bolivian regime included political instability; intramilitary conflicts, the total inability to resolve the economic crisis even to a minimal degree; and the scandalous corruption in the inner circles of the armed forces, clearly involved in drug trafficking, which grew three-fold during the military regime. Further, military diplomacy could not fulfill the promises it had made regarding economic and commercial cooperation, based fundamentally on the sale of Argentine manufactured goods to Bolivia and the purchase of Bolivian gas for an extended period. Undoubtedly, the country's deindustrialization, produced by the economic model (pushed by economic diplomacy), acted in a dysfunctional manner with respect to the expansionist plans of military diplomacy. As García Lupo has appropriately noted (1983, 36):

> This is why the attempts of the Argentine military to play a role in internal Bolivian matters has become more and more obsolete, and is based solely on the doctrine of national security, under which the ideological frontier of the two countries must be defended at any price. In the last few years that "any price" has meant the financial support of the Bolivian regime with clandestine resources, which for the most part, ended up in the pockets of the temporary military leaders and which were not able to impede to any degree what is now so evident: that Bolivia is mired in economic chaos and political unrest.

In effect, the growing and sustained deterioration of the internal Bolivian situation produced the disintegration of the process headed by García Meza. Thus, in October 1982, when Siles Suazo came to the presidency (his electoral victory of 1980 having been ratified by the Bolivian national congress), relations between the Argentine military regime and the democratic government of Bolivia entered a phase of stagnation.

Finally, the rapprochement with Brazil, beginning with the signing of the Tripartite Agreement on the Corpus-Itaipú hydroelectric projects between Argentina, Brazil, and Paraguay (October 19, 1979), was, to a certain extent, an exception to the inconsistencies and ambiguities of the foreign policy of the Process. In this case, recognition of the disparities of existing power between the two countries led Argentine military diplomacy to adopt an exemplary pragmatism that earned the approval of economic diplomacy. It is necessary to bear in mind that since the beginning of the 1970s, the "economic miracle" of Brazil and the stagnation of Argentina contributed to setting aside the old geopolitical scenarios of bilateral rivalry in favor of concepts that emphasized the danger of Brazilian hegemony over a weak and indecisive Argentina.[11]

Within this framework, the thesis began to take shape in diplomatic and military circles in Argentina that continuous competition would place

Argentina in the irreversible situation of being losers (Selcher 1985). Using this perspective as a point of departure, the diplomacy of the Process gave priority to cooperation in order to take advantage of the possibilities offered by the bilateral Argentine-Brazilian understanding. This would reduce the importance of conflict possibilities that had operated against Argentina given the different attributes of power in the two countries.

To be sure, after going through a first phase of tension due to the problem of the hydroelectric dams, the bilateral agreements (signed starting in 1979) opened up interesting possibilities for advances in the field of cooperation. Two days before the signing of the Tripartite Agreement, the Argentine foreign ministry released a communique describing this agreement as "excellent, from the energy viewpoint as well as from the political. The excellence of the agreement stems from the normalization and improvement of Argentine relations with Brazil and Paraguay."[12]

Despite the optimism of this document, the growing political and economic difficulties the Process was facing in that period made it difficult to take significant steps on stable foundations in the relationship with Brazil. In part, the economic policy designed by Martinez de Hoz widened the existing gap between the two countries and consolidated the status of Argentina as the "junior partner" in the relationship. Furthermore, in the Brazilian case, military diplomacy also relied extemporaneously on the security theme at a time when that particular tune had not resonated in Brasília for several years.

In this sense, the speech delivered by Videla thanking Figueiredo for a reception in his honor during his trip to Brazil in August 1980 was a clear example of the limited vision of the Argentine strategy. Referring to the international situation, the former president stated that

> It is unacceptable that those who are the objects of Marxist attacks not adopt the best and most effective measures to counter those attacks. We must take those actions which free will permits and which are the reason for existence of our way of life.

And he added:

> In order to make this possible, and to ensure that cohesion not be a cover for mere rhetoric, it is essential that all countries, acknowledging the differences in identity which exist between them, should join together with the same opportunities to debate the strategies and policies to be placed into practice.[13]

It is clear that Videla's ideas on establishing unspecified joint mechanisms for the development of combined actions in Latin America were not exactly aimed at the discussion of bilateral or multilateral strategies leading to an increase in the region's capacity for external action.

In effect, the relationship with Brazil could not be separated from the basic assumptions on which military diplomacy built its Latin American policies. Ultimately, the invasion of the Malvinas revived in Itamarati the old stereotypes regarding the unpredictability and volatility of Argentine foreign actions. Thus, after the Argentine military defeat, the Brazilian government assumed a cautious and well-founded policy of "wait and see" with regard to putting into effect different initiatives for bilateral cooperation. This task would also have to await the return of democracy.

THE ALFONSIN GOVERNMENT: A CHANGE OF DIRECTION

Argentina's return to a close relationship with Latin America, which occurred after the reestablishment of democracy, shows an important break with the regional policies of the military regime. With the advent of democracy, then, the Latin American policy of the Alfonsín government centered on the following objectives: strengthening of peace and discouragement of all types of arms races in the region, opposing any doctrine that subordinated the interests of Latin America to the strategic objectives of superpower conflicts, consolidating representative government on the continent, promoting policies for "regionalizing problems and their solutions," and pushing for Latin American integration.

With regard to the objective of strengthening peace, the democratic government developed a very active policy, which was most clearly seen in the signing of the Treaty of Peace and Friendship with Chile, in the support for a negotiated solution to the Central American crisis, and in the decision to modify the traditional conflict hypotheses with neighboring countries.

Shortly after Alfonsín assumed the presidency, leaders from Argentina and Chile met at the Vatican (on February 23, 1984) to sign the Joint Declaration of Peace and Friendship. By this means, they confirmed their will to solve the southern question as soon as possible, with the help of papal mediation. It is important to stress that this commitment goes beyond the Beagle dispute, since in the Declaration both governments expressed their decision to solve, through strictly peaceful means, any controversy that might arise between Argentina and Chile.

For the Argentine foreign ministry, the basic objectives of the negotiation could be summarized in the following points: establishment of the limits of maritime sovereignty; Chile's recognition of the bioceanic principle; and establishment of a system for pacific solution of controversies.

For his part, during a trip to the Beagle Channel area in May of 1984, President Alfonsín reaffirmed the need to increase integration with Chile, within the framework of national sovereignty. He indicated that "Chile . . . must have an outlet to the Atlantic and we to the Pacific."[14]

Along these same lines, the majority of Argentine political parties signed an Act of Coincidences (June 5, 1984), which states in its eleventh point:

The acceptance of the suggestions and proposals of Pope John Paul II regarding the Beagle—as a framework for peace—will permit the peaceful and definitive solution of the border litigation with the sister nation of Chile, will confirm the bioceanic principle and our rights with regard to Antarctica, and will begin the consideration and approval of numerous plans for social and economic complementarity between our two nations.[15]

After intense negotiations, in Vatican City on November 29, 1984, Argentina and Chile signed the Treaty of Peace and Friendship. This signing was preceded on November 25 by a voluntary plebiscite with the purpose of allowing the Argentine citizenry to express itself in favor or against the terms of the Treaty, made public by the Alfonsín government on October 20. The participation of eligible voters reached 70.07 percent, and the vote broke down as follows: 81.32 percent in favor of signing the agreement, 17.09 percent against, and 1.09 percent blank votes.[16]

Among its more important provisions, the Treaty grants Chile all the islands in dispute (Picton, Lennox, Nueva, and nine smaller islands); it fixes at three miles the width of Chile's territorial sea; it sets the limits of an exclusive economic zone; it validates the borders established by the General Border Treaty of 1881 (which provides for the perpetual neutralization of the Strait of Magellan and its open navigation to ships of all flags); and it rejects the direct or indirect use of force to solve any future controversies. To this end, it establishes, in the first instance, a mechanism for bilateral negotiation (Conciliation Commission), and it provides, in the second instance, for recourse to an Arbitral Tribunal whose decisions would be obligatory for the parties. It also creates a permanent Binational Commission with the purpose of increasing economic cooperation and physical integration between the two countries.

The treaty is a transaction, and creates a complex and definitive solution to the matters under litigation, as expressed in Article 14: "The limits stated are a definitive and immovable boundary between the sovereignties of the Argentine Republic and the Republic of Chile."

The promulgation of the document by the Argentine executive power on March 26, 1985, was preceded by a long and intense debate in the senate at the beginning of the same month. This debate ended with the approval of the treaty by a close margin: twenty-three votes in favor, twenty-two against, and one abstention.[17]

For its part, the Chilean military junta unanimously approved the treaty on April 11, and Argentina and Chile exchanged ratification instruments at the Vatican on May 2. From this moment on, the two countries held various meetings aimed at fostering the bilateral integration policies provided for in the agreement. These led to the official establishment, in October 1985, of the Binational Commission for Economic Cooperation and Physical Integration between Chile and Argentina.

To be sure, Argentine-Chilean relations include other topics and difficulties beside border problems. The present deep differences between the political regimes of the two countries (as in the case of Paraguay) have been an obstacle for rapid rapprochement on more permanent bases. Further, the close link between the internal political situation of the Pinochet government and its international commitments make Chilean diplomacy a relatively sterile terrain for any planned Latin American integration. In this context, the solution of the Austral problem will depend to a large degree, on the political future of the country beyond the Andes.

For its part, Argentina along with other Latin American and African nations, cosponsored the Brazilian proposal to create a zone of peace and cooperation in the South Atlantic. This proposal was adopted as a resolution of the United Nations in October 1986, and is in the framework of a policy that attempts to remove Latin America from superpower conflict and tries to eliminate the presence of nuclear weapons in the region.

In effect, this proposal is contrary to that of the frustrated SATO (South Atlantic Treaty Organization), which still evokes nostalgia in certain Southern Cone military circles, especially in Argentina. The proposal attempts to create a specific security system based on the recognition of the South Atlantic's own identity and the need to keep it isolated from East-West tensions.

From this perspective, the situation created by the Malvinas dispute is a real threat to the common Argentine, Brazilian, and Uruguayan interest in avoiding the geographical proliferation of nuclear weapons in the area. It is clear to these governments that an eventual sharpening of the Argentine-British crisis could lead to an East-West confrontation in the region. This is because the Soviet Union will not remain passive as it observes the growing militarization of the region by a NATO power, but rather will attempt to counter this process by means of equivalent policies (for example, the increase of its military capabilities in Angola and establishment of bases in that country). At the same time, it is necessary to include in this analysis the effects of increased tensions in the area due to commercial interests of these three countries. In the Argentine case, the figures are eloquent: of the 92 percent of its foreign trade which goes by sea, 75 percent travels on the sea lane along the Brazilian coast, 18 percent heads to the Cape of Good Hope route, and 7 percent passes through the Strait of Magellan.[18]

With regard to the objective of consolidating representative forms of government in Latin America, the interest of the Alfonsín administration in strengthening democracy has two aspects. First, to seek external reassurances for support of the process of internal consolidation of democracy, and, second, to advance the process of regional integration. According to President Alfonsín,

We respect the internal situation of each one of our countries, but we cannot avoid stressing our profound conviction that integration will definitively triumph only if it is based on democracy and popular participation.[19]

SELECTIVE INTEGRATIONISM

From the moment it assumed power, the Alfonsín government made important contributions to the development of new forms of regional cooperation that, through the search for specific and timely accords, attempted to go beyond the inoperative and indiscriminate integrationism of past decades. At the same time, this "selective integrationism" tried to show the fallacy of the alternative of "individual salvation" in the framework of a policy of alignment. In this latter sense, Foreign Minister Caputo, during his trip to Brasília in mid-May 1984, made clear the position of the democratic government:

> Up to not long ago, the Latin American nations individually, and sometimes collectively, only projected their destiny toward the developed nations, and this for some years blocked the integration process. But now, we have begun to learn the importance of continental unity, because only by that means can we improve our possibilities and increase the independent decisionmaking power of our countries.[20]

In the multilateral arena, the Argentine Radical government has carried out a primary role in the formation of both the Cartagena Consensus and the Contadora Support Group. In the bilateral arena, we should note the progress achieved with the democratic countries of the area under consideration: Bolivia, Uruguay, and Brazil.

The relationship with Bolivia reached a high point with the November 1987 signing of the Act of Buenos Aires, which established a permanent system of bilateral consultation, thus institutionalizing a high-level political dialogue to deal with bilateral, regional, or global matters. At the same time, after a long negotiating process, the countries signed a Memorandum of Understanding. An economic agreement, it provided for an overall solution to the issue of Bolivian gas imported by Argentina and the refinancing of the bilateral debt.[21]

As to Uruguay, in addition to the intense and fluid bilateral policies implemented when democracy was reestablished in both countries, the so-called Colonia agreements (May 19, 1985) created an Argentine-Uruguayan ministerial council for coordination and consultation, with the goal of promoting economic and social integration between the two countries.

With the signing of these agreements, bilateral economic integration began to grow significantly. In effect, bilateral trade acquired increasing importance: thus, Uruguayan exports to the Argentine market in 1986 reached 88.7 million dollars, a figure that exceeds by 40.6 percent the 1985 total (63.1 million dollars). With regard to Uruguayan imports of Argentine products, we should stress that Argentine sales to Uruguay during 1986 (118 million dollars) were 36.5 percent higher than those of the year before (86.4 million dollars).

Lastly, the clearest example of the selective rapprochement policy toward the region is the process of integration begun with Brazil in July 1986, which found its greatest thrust in the simultaneous transition to democracy in the two countries. This accelerated the tendency to draw closer together, beginning with the 1979 bilateral initiatives. It was strengthened by a series of important steps, among which were the Brazilian diplomatic initiatives in support of Argentine rights in the Malvinas and the coordination of positions in fora such as the General Assembly of the United Nations, UNCTAD, and GATT. By the same token, the debt crisis, commercial protectionism of the industrialized world, and the need to isolate the region from the East-West conflict all operated as cohesive elements that rescued (especially from the Argentine side) the old idea of the "pulling" capacity of the Argentine-Brazilian partnership in terms of individual and collective autonomy.

From the official Argentine perspective, the integration with Brazil sought the consolidation of the democratic process in both countries, the qualitative modification of politico-strategic and commercial bilateral relations, the strengthening and broadening of conditions for modernizing the country, and the defense of common interests in external affairs, both in political and economic domains.

As might be expected, the integrationist thrust of the government has aroused suspicions and resistance from some internal sectors. In this sense it is important to note that the greatest objections to the agreements came out of the economic sphere, and pushed traditional geopolitical questions to a second level.

In effect, the arguments made by some entrepreneurial sectors stressed the problems stemming from the different levels and degrees of development achieved by Argentina and Brazil. They also stressed existing doubts and fears stemming from financial and fiscal restraints under which the Argentine economy operates, and the differences between the two countries in matters of financing, regulations, and labor costs. These doubts, which are certainly legitimate in many cases, are to a large degree an expression of the corporate interests of a good many Argentine entrepreneurs, who are used to operating under the protective shadow of the state or in the framework of various plans for sectorial or regional promotion through which they have traditionally covered their operating deficits.

Despite these criticisms, the progress of negotiations between the business sectors and the government has been accompanied by a growing articulation, and this appears to indicate a greater support of the integration agreements by most of the Argentine entrepreneurs. The support of the political parties has been almost unanimous. Finally, despite existing suspicions in important sectors of the armed forces in both countries, the strategic requirements of the new Argentine-Brazilian bilateral relationship—as well as those emerging from the Latin American policies of Argentina and Brazil—appear to strengthen the development of convergence hypotheses to the detriment of the century-old conflict hypothesis that gave rise to the well-known "armed peace" in South America.

In this latter sense, it is important to mention the recent agreements in matters of nuclear, technical, and industrial cooperation. These are a step of enormous importance in the bilateral relationship, since the armed forces of both countries have traditionally been reluctant to carry out joint ventures of this type. Of similar weight was the July 1987 visit of President Sarney, accompanied by Alfonsín, to the Pilcaniyeu nuclear plant located in southern Argentina. The Argentine foreign office characterized the visit as a "maximum demonstration of mutual confidence between the two countries.[22]

In conclusion, there is no doubt that the simultaneity of the transition to democracy in the two countries has made possible the beginning of a process of profound reformulation of the bilateral relationship in cooperative terms. It is also clear that democratic continuity in Argentina and Brazil is a requisite in order to progress along the established path and definitively overcome the nineteenth-century nationalisms, the old rivalries, and the hegemonic temptations. Under the present circumstances, the maintenance of this course would not be merely a return to the postponed ideals of union, but rather the result of the palpable existence of common problems and interests that orient Argentine and Brazilian policies—as well as those of other Latin American countries—toward joint and concerted paths as yet unexplored.

NOTES

1. A good analysis of the objectives and composition of these new hegemonic coalitions formed under the shelter of the authoritarian regimes can be seen in Osvaldo Sunkel, *América Latina y la Crisis Económica Internacional*, Buenos Aires: Grupo Editor Latinoamericano, 1985.

2. With regard to the conflicts within government alliances during the latter phase of the implanting of the bureaucratic-authoritarian States, see the work of Guillermo O'Donnell, *Tensiones en el Estado Burocrático-Autoritario y la Cuestión de la Democracia* (Documento CEDES), Buenos Aires: CEDES, April 1978.

3. During the whole period of the Process, the Ministry of Foreign Affairs had an absolutely marginal role. In matters considered "key," it was left out of the decisionmaking process, and its role was limited to the implementation of policies designed in other spheres of government.

4. A detailed analysis of the guiding criteria of Alfonsín's foreign policy can be seen in my article, "Democracia y Política Exterior," in Mónica Hirst and Roberto Russell, "Democracia y Política Exterior: Los casos de Argentina

y Brasil," *Serie Documentos e Informes de Investigación,* 55, Buenos Aires: FLACSO/PBA, August 1987.

5. Regarding the different nuances of "Western" interventionism of military diplomacy (including parallel military diplomacy), see my article, "Las Relaciones Argentina–Estados Unidos: Del 'Alineamiento Heterodoxo' a la 'Recomposición Madura,'" in Mónica Hirst, ed., *Continuidad y Cambio en las Relaciones Estados–Unidos América Latina,* Buenos Aires: Grupo Editor Latinoamericano, 1987.

6. It is necessary to recognize that using a systematic propaganda action, accompanied by various mechanisms for self-censorship (which had been in operation from the beginning of the Process), the military regime was able to spread its geopolitical schemes—unfortunately, with considerable success—over a large sector of the population, and awoke in it strong anti-Chilean and bellicose feelings.

7. *La Nación,* January 26, 1979.

8. In this document the Vatican's diplomats validate the arbitration award's demarcation line, granting Chile the islands in question (Picton, Lennox, and Nueva), and all the other islands found to the South down to Cape Horn, drawing an enclosing line that covers all the lines in dispute. Using this line as a starting point, Chile has a strip of territorial waters of six miles, and a second strip of exclusive economic zone of another six miles. The remainder, out to 200 miles, will be devoted to common or joint activities (Sea of Peace), with Argentine sovereignty, but at the perpetual service of Chile.

9. The Carter administration, indignant over the involvement of Argentine military diplomacy in Bolivia, suspended the programmed visit to Buenos Aires of Assistant Secretary of State for Hemispheric Affairs William Bowdler, and expressed to the military junta its concern over the complicity of the Argentine military with the Bolivian coup-makers, and for the "deviations of the excesses." By way of reprisal, it decided not to appoint an ambassador to Buenos Aires until the U.S. presidential election in November of 1980, and the dialogue between the two countries was in effect interrupted until Ronald Reagan took office.

10. *Clarín,* August 6, 1980.

11. See, among others, Pablo Sanz, *El Espacio Argentino,* Buenos Aires: Pleamar, 1976. In this work, there are references to Brazilian strategy as a "geopolitical plan which tends to round out the sphere of her dominance. Brazil's efforts cannot help but wipe out substantial aspects of the attempts by the other States of the southern part of the continent to integrate and develop regionally, at the same time as they damage the minimal security requirements which States are required to keep for themselves" (p. 335).

12. Published in *Estrategia,* 61 and 62, November/December 1979 and January/February 1980.

13. *Clarín,* August 20, 1980.

14. *La Nación,* May 6, 1984.

15. *Tiempo Argentino,* June 8, 1984.

16. It is interesting to note that this consultation—beyond discussions which surrounded its constitutionality and the commentaries of various opposition groups regarding the eventual political purposes pursued by the

Alfonsín government—became converted into an important instrument for the transition to more participatory and direct political modes, capable of shaking the profoundly unpleasant remnants of an authoritarian and parochial culture.

17. The affirmative vote included the Radical senators (except for the abstention of Luis León), in addition to the *bloquista*, Liberal, and Autonomist senators from Corrientes, the *desarrolista*, and one of the two *populares* from Neuquen, Jorge Solana. The other, Elias Sapag, opted for rejection, as did the twenty-one representatives of the Peronista group. With regard to the different positions taken on this issue, see my work, "El Diferendo Austral: Posiciones Divergentes," *América Latina/Internacional* 1 (2), 1984.

18. *Ambito Financiero*, July 17, 1986.

19. *Clarín*, March 20, 1984.

20. *Tiempo Argentino*, May 15, 1984.

21. With regard to the gas issue, agreement was reached on a formula for tri-monthly adjustments based on the price of the fuel in various markets and that, as a result, bring the price of the residual Bolivian gas to international levels. As far as the Bolivian financial debt to Argentina was concerned, it was refinanced for a period of twenty-five years, with an eight-year grace period, at an interest rate of 8 percent.

22. With regard to agreements on nuclear matters, and technical and industrial cooperation, see Gloria Fernández, "Argentina: las Relaciones Especiales Bilaterales," *América Latina/Internacional*, 5 (15), January/March 1988.

part three

BRAZIL

chapter five

The Southern Cone and the International Situation
Therezinha de Castro

In 1494, the Tordesillas treaty segmented South America geopolitically into two zones of maritime influence: the Pacific zone and the Atlantic zone. Consequently, from a geohistorical point of view, the area customarily called the Southern Cone, a subsystem of the American continent bounded by the twentieth degree of south latitude, would become almost totally enveloped in the Spanish zone of influence on the Pacific—a secondary position during the age of discovery. This historical fact would contribute to the discontinuous occupation of the Southern Cone, creating a region of dispute for the best position on the Atlantic and establishing zones of tension that still exist.

The Southern Cone is, geographically speaking, a smaller continental territory than the territory north of the twentieth parallel, because the South American continent widens to the north. Therefore, the Southern Cone can be called a subregion of the continent, with a characteristic maritime influence. Considering this maritime impact, then, there is geographical justification for having chosen the twentieth parallel as the northern limit of the Southern Cone:

- On the Pacific side, the Southern Cone comprises almost half of the South American coastline, all of which belongs to Chile, and it splits the Andes at the Bolivian highlands, exactly at the point where the mountain range begins forming such passes as those at Santa Rosa and Uspallata, which facilitate communication between the two oceanic watersheds.
- On the Atlantic side, the southern coast is characterized by a gradual curving to the west and becomes transformed into an integrated territory very much dependent on the oceans.

Translated by Sílvia Beatriz A. Becher Costa

Map 5.1   Southern Cone Position

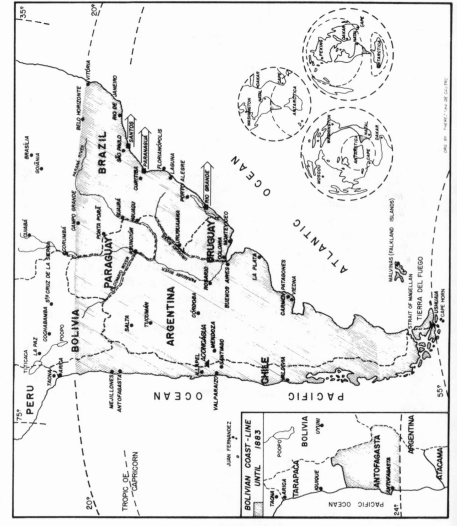

The disputes at the mouth of the La Plata, the natural entrance to the Southern Cone on the Atlantic, led Brazil, still a Portuguese dependency, to cross the limits set by the Treaty of Tordesillas and become involved in this continental subregion. In geopolitical terms, during this period of occupation and dispute, the Portuguese maintained a united front in their Brazilian colony. This was in contrast to the Spanish divisiveness or cantonalism projected in various administrative fronts and *audiencias*. The Southern Cone thus comprised two entities controlled by the Kingdom of Castile: the Captaincy General of Chile and the Viceroyalty of the Rio de La Plata—both of which had been separated from the Viceroyalty of Peru. The areas lining the Atlantic and Pacific Oceans, which were separated by an unsettled hinterland, would develop independently, thus emphasizing the isolation of the Captaincy General of Chile but not the Viceroyalty of La Plata, which had two ocean fronts. This geographical and historical individuality, reenforced by the Spanish geopolitical cantonalism and the physiographical characteristics of such a mountainous territory, would greatly contribute to the Balkanization of the Southern Cone. This explains why Brazil coexists in the Southern Cone with five Spanish-speaking countries—Chile, Argentina, Uruguay, Paraguay, and Bolivia.

The geopolitical entities formed by the Viceroyalty of La Plata and the Captaincy General of Chile had historically been secondary territories for the Castilian Ultramarine Empire, since the southern part of the continent (the Southern Cone) would receive less importance under the mining mercantile spirit of the Viceroyalty of Peru, which had also been separated from the Pacific by the strong attraction of the North Atlantic, via the Isthmus of Panama.

This secondary position also came about because, as the Southern Cone was dependent on the Pacific Ocean, Spanish influence was much more concentrated to the north along the Peru–New Hebrides and Mexico-Philippines axes. Furthermore, this neglected part of the Pacific triangle was secluded from the main commercial routes of the Southern Cone—those which depended on the Atlantic and were then flourishing on the Plata estuary.

MARITIME GEOPOLITICAL UNITS

Exclusion from maritime mobility was a disadvantage for the Southern Cone during the early Iberian colonization period. This negative factor was made even worse because the Strait of Magellan was considered a less promising route to the Indies, the spice lands, than was the route along the Cape of Good Hope.

The primary concern with the Indies would induce settlement and development of the Southern Cone, which would be effectively carried out on the Atlantic coastline. This preference for the Atlantic, where the Portuguese and Spanish confronted each other, would, starting in 1680, transform the mouth of the La Plata into a zone of dynamic struggle of domination for this hydrographic basin, a 3,200,00-square-kilometer region that encompasses one-third of South America.

An analysis of this geopolitical disjunction shows that, while the La Plata viceroyalty was involved in competition over possession of the La Plata Basin, the Captaincy General of Chile assumed responsibility for exploring the Patagonian trans-Andean area. Because Patagonia (which today is part of Argentine domain) was then deprived of maritime mobility, its population-growth rate was low, resulting in its geopolitical neutrality. This happened since Buenos Aires was, for some time, becoming an authentic terminal for South Atlantic routes, thus leaving practically 80 percent of its coastline isolated from international trade. In order to lessen this geopolitical focus on the city today, the Argentine government plans to shift to a new capital city in the south, in the proximity of Viedma-Carmen de Patagonia.

An identical phenomenon occurs in the Pacific area, where the Southern Cone includes only Chile, its territory extending some 4,200 kilometers (which corresponds to a north-south distance from the peninsulas of Scandinavia to Italy). A narrow strip of land compressed by the Andes and the Pacific, Chilean territory is widest (350 kilometers) near Mejillones, while it tapers to a mere 100 kilometers near Illapel. Located between the sea and the mountains, Chile's communications with its neighbors are not easy: it is necessary to cross lands above 4,000 meters in altitude to enter Bolivia or Peru. Because of its unique geographic characteristics, the country's active nucleus is located in the Central Valley, along the Santiago-Valparaiso axis, where ninety percent of the population lives. This nucleus, nevertheless, depends on the Atlantic, reaching it either by means of the transcontinental railway to Buenos Aires or by means of the passes of Valle Hermoso, La Cumbre, Los Puigenes, and Las Leñas.

It may be concluded, then, that the long and insurmountable maritime distances are highly unfavorable for Chile and Argentine Patagonia, in contrast to the activities along the Brazilian coast. In the Southern Cone— the zone of highest Atlantic-Pacific dependency—the maritime mentality was most markedly present in Brazil, even though Chile, Brazil's opposite in this regional subsystem, might be supposed to have a greater maritime thrust because of the oceanic determinism of its position. In fact, in Chile, where the distance from north to south has a disintegrative effect, the unity of the Central Valley could compensate for the striking differences found in the Chilean shape.

Uruguay, although a more geographically compact country with relatively shallow coastal waters, is likewise detached from a maritime mentality, maintaining in Montevideo its predominant center not only in geoeconomic terms, but also in geohistorical, geopolitical, and geostrategic terms. The establishment of Montevideo (1726) as a strong Spanish fortification was a geostrategic move to impede Portuguese expansion at the colony at Sacramento, an inland Portuguese position in the La Plata region. On the other hand, to prevent Montevideo from dominating the larger hinterland, the Portuguese government ordered the settlement of Rio Grande de São Pedro (today corresponding to the state of Rio Grande do Sul, on the southern end of Brazil) from 1737 onward.

From a geopolitical standpoint, therefore, the desire to dominate the La Plata estuary, a zone of vital importance to the Southern cone, would seem to have triggered the struggle that has lasted almost two centuries in Uruguay and to have speeded up settlement in southern Brazil beyond the Tordesillas limits.

THE GEOPOLITICAL POLE OF LA PLATA

The crucial phase of these disputes ceased by 1870 and guaranteed free navigation in the La Plata Basin benefiting the ports both of Buenos Aires and Montevideo. These ports profited from their position as geopolitical poles of the Southern Cone, not only in the physio-political aspect of their natural north-south axial location, but also because of the attraction of the La Plata estuary, which led to the inland countries of Paraguay and Bolivia, both relying upon Argentina and Uruguay for a direct outlet to the Atlantic.

Later on, considering that the key to controlling the Basin lay in Paraguay, it would be possible to establish a more direct sea outlet for the two landlocked nations by crossing Brazil to gain access to Atlantic ports. This contact was, in fact, later set up, as both Brazil and Argentina were involved in four of the seven binational frontiers coexisting within the Basin, which assures Brazil a constant presence in La Plata watershed affairs. Thus, to oppose the attractiveness of the concentric axes of Buenos Aires, a set of east-west routes was constructed, including both railroad and highway systems, which would rival the Argentine system of communications with the interior.

In order to counterbalance the exclusive geocentrism of the La Plata estuary, it was considered necessary to establish several other Brazilian outlets, thus eliminating the isolation of the La Plata interior. This resulted in the Export Corridors policy, a policy effectively developed from 1960 onward, when Brazil adopted the philosophy that "exporting is a national need." Coincidentally, the first exporting terminals were found mainly in the La Plata Basin, as these offered more adequate facilities and were located at the confluence of highways, railways, and waterways. They served agricultural centers as well as the Brazilian multi-industrial poles of Porto Alegre/Rio Grande, Curitiba-Paranaguá, and Rio de Janeiro/São Paulo-Santos, and they attracted in addition the hinterland centers of Corumbá/Campo Grande and Brasília/Goiânia.

The export corridors are, therefore, part of Brazil's policy of integration, and these efforts are greater in the La Plata Basin because of its importance within the Southern Cone as the most effective geopolitical pole encompassed by the entire South Atlantic. Under such an integration policy for the interior La Plata (the main door to the Southern Cone), however, Brazil must conform to the contrasts of the region. On the one side, such contrasts reveal Paraguay, with its most densely populated and economically developed area turned toward Brazil and with its connection to Brazil made easier because of adjacent plains, both areas totally integrated within the La Plata, similar to an authentic Mesopotamia. On the other side, there is Bolivia, with its main ecumene located in the highlands. More dependent on the Pacific through its Cochabamba-Arica linkage, Bolivia has its less important geopolitical areas turned toward Brazil, therefore indicating that Bolivia, in contrast to Paraguay, has a smaller attraction to the Atlantic.

The Santos, Paranaguá, and Rio Grande export corridors not only have allowed for the integration process to occur, but also have given the La Plata Basin greater geostrategic amplitude in South Atlantic concerns. Offering several port options now, the region cannot, in the case of an armed conflict, be as easily blockaded as in the past, when it could only count on the Buenos Aires–Montevideo outlets.

Of the 120 million inhabitants who make up Brazil's total population, according to the 1980 census, it is noteworthy that the La Plata area alone holds 50 million Brazilians. Consequently, even when one considers the combined Spanish-speaking population for the La Plata countries, the Brazilian population in the region is still larger. This comparison holds true likewise when one examines the total La Plata area, for Brazilian territory occupies more than half of the combined Spanish territories in the basin.

Not only does Brazil stand out because of its critical mass (a combination of factors including its area and population), it also plays the role of a multiple vector toward the Southern Cone's geopolitical pole, owing to its position of extensive coastline in terms of the basin and in terms of its continental mass, which is similar to those of North America, Europe, and Africa.

Therefore, in discussing regionalization in South America measured according to pacts and systems, the Treaty of the La Plata Basin requires a special focus. With the purpose of promoting harmonic development and physical integration within this geopolitical pole of the Southern Cone and its "direct and probable areas of influence," the regional accomplishments to date can be summarized thus: the rational use of water resources (Itaipú being the largest hydroelectric complex in the world), the economic advances in frontier areas, and the improvement of highways, railways, airways, and electrical and telecommunications systems for achieving better integration of the area.

As it does not possess territory in the La Plata Basin, Chile remains the only Southern Cone country not participating in the treaty, signed in Brasília on April 22 and 23, 1969. Because of the importance it places on linking itself to the other nations of the Southern Cone—after all, the region contains the country's primary ecumene and the vital geopolitical triangle of Rio de Janeiro–São Paulo–Belo Horizonte—Brazil's foreign affairs office, the Itamaraty, gives special attention to its diplomatic activities in the region. It endeavors, consequently, to consolidate confidence, understanding, and cooperation among all partners of the Southern Cone.

The reason for this attitude is that the Southern Cone constitutes the strategic appendix of the continent, due to its wedge-shaped end on the Atlantic-Pacific terminal. Although Brazilian territory does not reach the southernmost region of intricate channels and straits, Brazil's participation is essential, both because of the active control it exerts over the Atlantic coast and because of the dynamic attraction it maintains within the La Plata Basin.

INSTRUMENTS AND ZONES OF PROJECTION

Although we can observe different map projections of the Southern Cone, we still will always see a constant in the region's contiguous zone—the South American Antarctic with sub-Antarctic archipelagoes, South Africa, and, in a more remote position, Australia. And within this general setting, the Southern Cone stands out as a wedge, the continent's base in the south comprising the Drake Strait and the Straits of Magellan and Beagle, with territorial waters that offer the most intricate and varied communications between the Atlantic and the Pacific Oceans the Rio Pact encompasses.

The Inter-American Treaty of Reciprocal Assistance (Rio Pact)

Signed in the city of Rio de Janeiro on September 2, 1947, and known as the Rio Pact, this document was, for some people, just one among many other episodes that the inter-American system had already grown accustomed to. But, for others, the treaty constituted the first instrument of legitimate collective defense, based upon Article 51 (Paragraph 3, Section 1) of the United Nations Charter. The treaty, then, is grounded on the recognition of the right of legitimate defense, associated with a continental solidarity that the United Nations Charter supports and recognizes as a valid regional system or agreement.

At first, mutual defense against external aggression was directed against the Axis forces, although shortly after this it came to a new position against Soviet expansionism. Mutual defense has never worked in practical application because it involves countries with frontier disputes or contested sovereignties extending from Central America to the Southern Cone and even involving sub-Antarctic archipelagoes and Antarctica itself; it was seldom applied to the peaceful settlement for which it was intended.

The treaty had not been invoked for settling internal disputes since the 1950s, and the Malvinas crisis in 1982 revealed in practice that it continued to fulfill Washington's objectives for consolidating an anti-Soviet front in America.

As a whole, considering its practical defensive goal, the Rio Pact has never been successful, for the United States, being committed to NATO since 1949 and defending its national territory directly, chose instead to deal with security through bilateral agreements. Hence, the establishment of military and technical organization models, as gradually set up by the United States throughout the continent from Mexico to the Southern Cone, succeeded for a while in creating a friendly, safe, and stable front for the United States in South America. For this reason, relying on inter-American military missions as provided through the Act of Chapultepec, the United States geostrategically diminished the importance of the Rio Pact, using it to plan for actions only in cases of immediate aggression or threat of aggression.

The continent the Rio Pact attempts to cover as its defense zone is an authentic island, the security of which is directly related to freedom of navigation on the Pacific and Atlantic oceans. These oceans are not totally covered from east to west or from west to east by the pact. Thus, without a regular armed alliance such as NATO, or without an obligatory support, the Rio Pact can be called inter-American, but it is in no condition to honor any mutual assistance request. While NATO covers all of the northern flank of the continent from Alaska to the Canadian arctic, the Rio Pact has patently failed to operate on the two interconnecting areas of the Atlantic-Pacific: the Caribbean and the Southern Cone.

Although the United States maintains logistical support bases on its "fourth frontier" of Central America, an analysis of the terms of the Rio Pact indicates that a vast portion of coastline extending from the Caribbean to the Antarctic is left unprotected. In this zone, Brazil can be considered the key point—both in terms of its location and in terms of its political line of action—in the tension zones extending from the "fourth frontier" to the Malvinas. Brazil not only belongs to the "Caribbean Margin," as established by the Amazonian Pact, but it also possesses the less protected and vulnerable area of the South Atlantic via the Natal–Dakar Strait to the Southern Cone. Without Brazil, the entire austral hemisphere becomes vulnerable, particularly when we consider that the northern and western regions of South America are potentially as unstable as is the Caribbean. Therefore, according to some analysts, to include Brazil in a line of defense means to assemble a coastline from Miami to Antarctica secured against any attempt to launch guerrilla action in the Andean nations.

In general, however, and considering its national and international integrationist programs (particularly since the 1960 inauguration of Brasília) Brazil has become a nation that maintains cordial relations with its American neighbors and simply coexists with enemies that might come from overseas.

It is the United States, then, in case of deadlock, that is expected to assume leadership of a reliable geostrategic coalition capable of obtaining the cooperation of Rio Pact members. And because the American continent is, in fact, an island, joint action in the central regions of America would have to rely on mutual support in the geostrategies of both the North and the South. Thus, the Inter-American Treaty of Reciprocal Assistance, in view of the present antagonistic world, can be regarded as totally obsolete in terms of collective security.

Antarctica: Its Implications

Whereas Articles 1, 2, and 3 of the Rio Pact deal with immediate reaction by individual or collective means, as well as with coordination of measures to be taken in cases of threats to continental peace, Article 4 indicates the limits of the treaty's geographic zone of action from one pole to another.

Initially, the security zone proposed in Article 4 followed a restricted demarcation, with points of support in Alaska and Greenland, leaving out the Southern Hemisphere areas of Antarctica and the sub-Antarctic archipelagoes facing the Southern Cone later incorporated by Argentine imposition. The pact, which has its core in the Southern Cone, at best only included the northern area of the Drake passage, even though the International Hydrographic Bureau has always considered glacial Antarctica as an authentic extension of the Atlantic, Pacific, and Indian oceans. Thus, by the impact of cold currents—either the Malvinas current in the Atlantic or the Humboldt in the Pacific—there is an intimate connection between the Southern Cone nations and the Antarctic.

The Department of State in Washington has two different policies toward the Antarctic: a "territorialist" policy and an "internationalist" policy. The former is in favor of demanding a political partition of Antarctica, claiming territories already staked out by Byrd and Ellsworth and situated between meridians 90 degrees and 150 degrees west. The internationalist policy favors internationalization of the continent. This latter policy, nevertheless, opposes two U.S. principles pertaining to the region: the U.S. Air Force's thesis that Antarctica possesses strategic value, and the U.S. pledge to abide by Article 4 of the Rio Pact, which accepts the South American Antarctic as a zone that directly affects the Southern Cone.

The tendency among nations to establish bases of support in Antarctica shows that the continent serves not only as an important early warning location but also as an essential interception and emergency base for Southern Cone and other nations affected by South Atlantic–South Pacific defense. And if, in practice, it has been proven that the Arctic serves both peaceful and military goals, the same application can be made concerning the Antarctic, especially when we consider that the world political axis has been gradually moving from East-West to North-South. Under these conditions and because of the ineffectiveness of the Rio Pact, NATO cannot reach this choke point in the South Atlantic, its bounds not extending below Gibraltar. Indeed, as a simple reaction to Soviet policy in Europe, NATO does not reveal a solid front toward the American continent, allowing a vulnerable Southern Cone flank to remain unprotected.

By setting such a geopolitical limit it imagined to be geostrategic, and in an historic agreement that ignored geography, NATO members did not even consider that during World War I, not all battles were fought north of the Tropic of Cancer. Moreover, they did not consider the Battle of the South Atlantic during World War II, when a considerable tonnage of shipping was lost to the "maritime guerrillas" of Axis battleships and armed merchant vessels. The electronic war of the Malvinas conflict in 1982 not withstanding, the geostrategic realities have not changed very much. And in the case of the Atlantic, Pacific, and Indian Oceans, where the Southern Cone stands out as a wedge, the Kremlin takes advantage of Western bloc disagreements and, consequently, has been gaining ground, changing the area into a potentially significant zone of confrontation.

At present, in terms of local security interests, four Southern Cone countries, Brazil, Argentina, Uruguay, and Paraguay, are members of the South Atlantic Area Maritime Command, which is responsible in a limited way for coordinating plans and exercises for the protection of shipping in the area. Under these circumstances, on the Pacific side of the Southern Cone, an alliance favorable to the Western bloc would be even more feasible, since it would involve Chile exclusively. On the Atlantic side, however, despite the presence of Uruguay and Argentina, Brazil would dominate the ocean to such an extent that a close relationship with Brasília would make it unnecessary to establish further agreements with South American countries for gaining access to peripheral airfields that enable antisubmarine warfare to be conducted with land-based aircraft. This is because, from the moment a merchant vessel passes the Cape of Good Hope in Africa in transit through the Atlantic until it crosses 15 degrees north latitude on its way to New York, London, or Gibraltar, it will always be closer to Brazilian ports than to any ports of the other South American countries. This situation gives rise to the premise that to negotiate with Brazil for the use of bases will provide maximum flexibility for transferring aircraft from one location to another, which will allow a concentration of forces in any area of the convoy route as the case may require.

CONCLUSION

The Southern Cone geostrategically dominates Mackinder's External or Insular Crescent in a zone of projection remote from the East-West axis. Nevertheless, its location can still affect the security of U.S. territory. Until recently, this area was forgotten in terms of a potential theater of operation.

But this situation has been changing since Southern Africa, a projection zone of the Southern Cone, has been caught in the "Soviet tongs" as a consequence of the success of the Gorshkov Doctrine.

At this time, political *aberturas*, or democratic "openings," are emerging in Southern Cone nations, after a decade or more under military right-wing governments, such as still exist in Paraguay and Chile. The United States must remain on guard, especially since it is known that the Soviet Union is an unconditional defender (sometimes using indirect methods in situations where it is more advisable not to act in the open) of regimes that submit themselves to its protection, as was the case of Salvador Allende's Chile until September 1973. Whereas the Soviet Union tries, in a practical and effective way, to establish its position in the South Seas, there appears to be no unified reaction to this penetration nor homogenous evaluation of the factor in the Southern Cone because of the many diverse Western policies toward the area.

Standing as a bastion for the West's protected rearguard, the Malvinas crisis brought dissent to the Southern Cone. At the time, both the OAS and the Rio Pact revealed themselves to be much too symbolic and, thus, of little effectiveness given the absence of coordinated action. If any real consequences derived from this fact, it was the practical exclusion of the United States from Rio Pact membership when it instead shifted from political support to actual military aid in complying with its NATO commitments.

Another, still latent, area of conflict which, like the Malvinas, lies in the Southern Cone and might eventually involve the United States, is the area of the Province of Antofagasta, annexed by Chile in 1883. Although Chile in 1904 built a railway linking La Paz and Cochabamba in the highlands to the Pacific, thus opening the ports of Arica and Antofagasta to Bolivia, the succeeding Bolivian governments have continued to demand the return of its coastline.

Based on the issues discussed above, it becomes clear that South America and the Southern Cone were only included among U.S. priorities by accident. Considered primarily as an unexplored area within the geostrategies of bipolarity, this region has drawn Washington's attention only because of its globalist policy.

In the context of the cold war, there is undoubtedly an inclination in Latin America, and particularly in the Southern Cone, for the formation of "nationalist" regimes, supported by mass appeal, through exploiting the historic concept of U.S. "imperialism." Nevertheless, in spite of a certain amount of anti-U.S. sentiment to the United States, the positive presence of the West constitutes an obstacle to closer ties with the Communist bloc, preventing a regional plan of containment.

Consequently, it appears that the U.S. National Security Council does not feel there exists any danger of a communist attack against any Southern Cone nation, except in case of global war. However, this same National

Security Council is aware that the Communists stand a good chance of controlling the nations of the area via electoral methods, when they do not succeed in participating directly in national policymaking. Thus, in 1958, there were public demonstrations of ultranationalist groups in Brazil and Argentina. Indeed, the Operations Coordinating Committee report (1958) even criticized the Brazilian government's lack of interest in fighting against communist infiltration, although it gave direct credit to the rightist tendency of the military.

As security issues have worldwide consequences, they cannot be seen as exclusively regional. The defense of the hemisphere depends as much on the protection of the northern flank as the southern flank, since both provide access to local resources and secure communications with the rest of the world. Only through effective cooperation with the United States will security for the two flanks be maintained, preventing armed attacks on a specific area that might be affected by a crisis.

The 1958 report of the Operations Coordinating Committee states that the unique position of military groups in Latin America makes U.S. influence on them very important. A fundamental factor in maintaining this influence is to offer training and equipment. In some cases, the report continues, it is even more attractive for the United States to guarantee all of this equipment, principally for political motives.

Under the human rights policy, a landmark during the Carter administration, not only had this influence declined, but U.S. military assistance had also been scaled down, thus reducing the desire and capability of the southern flank countries to support Washington's interests. Although it is evident that the Southern Cone nations are unable to defend themselves against a well-organized armed opposition without receiving help from the leader of the Western bloc, this merely reflected the greater attention the Carter doctrine devoted to the West-East axis centered in the Middle East and the Persian Gulf.

Availing itself of the resulting power vacuum, Moscow continues in its efforts to infiltrate through diplomatic, economic, and even military means. Within the scope of the Southern Cone alone, for instance, Moscow has become an important consumer of Argentine grains and is Brazil's sixth most important commercial customer.

Maintaining the premise that mutual cooperation is always more difficult than mere coexistence, Brazil gradually came to realize that belonging to the Western bloc did not mean integrating itself into the First World; with this new awareness, the country started to follow a philosophy of "responsible pragmatism," especially after 1970, based on a mature relationship with the United States that did not imply an automatic alignment.

At the same time, the U.S. liberal establishment began pressing Congress to suspend military and economic assistance to some countries of the Southern Cone, especially Chile and Uruguay. Consequently, North-South relations became a system of intersecting lines; the intimacy between the liberal North and the conservative South became increasingly difficult, with the Southern Cone nations standing together on the national issue and voting in unison when any sanction against Chile was at stake, as occurred during the OAS meeting of 1977.

Distancing itself from the leader of the Western bloc, Brazil renounced its Military Assistance Agreement with the United States on March 11, 1977. In view of Brazil's decision, and having seen no substantial increase in its program of military cooperation intended to supply equipment for collective defense, the United States began to lose the southern flank, a leadership of a credible geostrategic coalition in the Southern Cone, with the capacity of continuing the Rio Pact in force.

In turn, Brazil, without political inhibitions and with the encouragement of its partners, started to expand and enforce its presence in the Southern Cone. And this autonomy of Brazil from the United States reflects, above all, the ambiguity of the United States, which has encouraged Brazil to take its new stance.

The United States' ambiguity was caused by the confusion of regionalism and globalism in its foreign policy—confusion that had its source in the institutional struggle between the U.S. Security Council and the Department of State. The former agency bases itself upon globalism, and the latter, because of its structural organization of departments with specialized staffs for certain world regions, prefers regionalism.

Both globalists and regionalists, however, ask themselves the same questions:

- What is the principal U.S. interest within a certain region?
- What is the nature and magnitude of the Soviet threat?
- What U.S. policies will best defend the country's interests in relation to the Soviet Union?

Nevertheless, what distinguishes the two policy currents are the different priorities assigned to these questions and the emphases placed on each answer. Whereas globalists highly value Soviet and other Great Powers' behavior, threats, and motivations, the regionalists begin by defining, above all, the interests of the United States in each region. Therefore, globalists accuse regionalists of not duly considering Moscow-Washington competitiveness, regionalists respond that, because a geostrategic perspective of the cold-war focus on regional problems within a socio-political and economic viewpoint has been ignored, the United States has lost its former position of influence in the Third World, where power vacuums have followed in the wake of successive waves of instability.

Whereas Franklin Delano Roosevelt established the regionalist Good Neighbor Policy in the continent without interfering in rightist dictatorships that had proliferated at the time, Jimmy Carter chose as a point of honor the globalist defense of human rights. But in reaction to this, waves of instability swept through the hemisphere, especially in the Southern Cone. It could be said that even the United Nations became opposed to the Carter policy, forcing the United States to turn to other audiences, while the passivity of the OAS to the Monroe Doctrine relegated that organization to a secondary plane.

Under Ronald Reagan, the former policies of John Foster Dulles have returned, through which the United States placed more emphasis upon attainment of defense interests than upon making friends in the hemisphere. And, considering its interests in the Strait of Magellan, the United States stood firmly at Great Britain's side in the Malvinas crisis, disregarding any intention of keeping its friends in the Southern Cone. In view of the Malvinas, therefore, it is to be concluded that U.S. globalism aims to reestablish the widely known "safety net" margin in military equilibrium with the Soviet Union, preferring a "stable" ally such as Great Britain over an "instable" ally such as Argentina.

During the Malvinas Crisis, Brazil showed that it does not participate in the global problems of the United States, for Brasília immediately recognized Argentine sovereignty over the archipelago, revealing an entirely South American regionalist commitment. Nevertheless, because of Brazil's presence and position, the policies of Washington for the region would not seem to be adequate without the tacit support of Brazil. Bilateral relations are important, although they do not represent a new convergence or identical interest on the global plane. They are important because, even in view of the greater independence of Brazilian policy, it is still necessary on a regional level for the country to coordinate its actions with the United States. The existence of a gap in bilateral relationships is to the benefit of neither of the two countries involved.

Brazil, as an emerging nation living in an adolescent portion of the geopolitical cycle, should not be seen as a mere representative of the United States. Considering world issues currently influencing Southern Cone affairs, Brazil ought to be attractive as a friendly country as well as one that is well positioned for playing an important role in the conflicts of the Atlantic region and in the turmoil of the Caribbean and of the vulnerable maritime routes on the West African coast to the Indian Ocean.

The Southern Cone is more isolated on the Pacific side, since it is a great distance from its projection zone, Australia. This geographic function places Brazil in the central area of Mackinder's External or Insular Crescent, as a part of the Southern Cone and possibly as a part of a bloc of states encompassing the South Atlantic–South Pacific Oceans, in both cases being in a central position that ensures Brazil's direct involvement with almost two-thirds of the member states of the United Nations.

Furthermore, it can be said that, in terms of its international situation, the Southern Cone is also inserted in the Pacific, an "isolated ocean" whose shipping is primarily regional. Nevertheless, the Southern Cone is circumscribed as well by the Atlantic, an ocean of intensive shipping and intercontinental circulation. As a consequence, the Atlantic zone reflects more clearly the complex interplay of influences of the Eastern and Western blocs, where NATO has its most unguarded flank and where the Gorshkov Doctrine has increasingly collected its greatest victories.

Considering the "geopolitics of confrontation," the roles played by the Soviet Union and the United States in terms of international bipolarity are quite affected by the positions these superpowers may occupy on the world map. Although aero-spacial power has begun to gain greater impact, the superpowers are still much more dependent on ocean spaces. Consequently, access to the seas is one of the most favorable positions of strength in the present geopolitics of confrontation. In the world chess game, therefore, a geopolitical position that allows for as much influence as possible over important communication routes, that is, Mahan's concepts of international choke points and bases, should be the goal of the two competing blocs. In the present international situation, a country's participation in the Southern Cone must be analyzed and assessed according to this premise. If, in the North, NATO controls ocean-entry points of the Soviet fleets, in the South, the Gorshkov Doctrine of maritime projection threatens the "soft underbelly" of the Western world.

From 1980 on, because of the proliferation and diffusion of world power, the geopolitics of confrontation has triggered formation not only of the East-West axis but also of the North-South axis. The latter is rather more complicated, given the heterogeneity of nations that compose it. It is worth pointing out, however, that in America, if the North-South axis cannot be considered homogeneous, there still is some consistency in its cultural origins, its Western civilization, and its capitalist geostrategies. Consequently, this same America, a continental island surrounded by both the Atlantic and the Pacific, with an appendix on the Southern Cone, would be seen by Haushofer as the only territory capable of setting a counterbalance to Eurasia-Africa.

chapter six

Brazilian Geopolitics and Foreign Policy
José Osvaldo de Meira Penna

Brazil is a product of diplomacy. Brazil's destiny was settled even before discovery (April 1500) by an act of international policy that decided that this territory, bathed by the mild waters of the South Atlantic, would belong to the Portuguese branch of Mediterranean culture, Latin and Catholic. Southey has claimed that Brazil was discovered by chance. We could submit the thesis that, on this occasion, the chance that presides over the birth of nations had a diplomatic instrument to foreshadow its decree. It was the Treaty of Tordesillas of 1494 that set the division of lands to be discovered in the New World between the Iberian monarchs, thus precluding a serious clash that both the Catholic kings and Dom João II of Portugal contemplated with misgivings. Tordesillas marked the direction of conquest. The course of navigation had already been anticipated by the expedition of Columbus and of Bartolomeu Dias. The treaty simply ratified the paths of discovery, that of the Spaniards westward—*el levante por el poniente*—and that of the Portuguese eastward, around the bulk of Africa. The destiny of Brazil was thereby also linked to that of Africa. The country was born out of an unfulfilled dream (of Dom Afonso V of Portugal) of imperial Iberian union and of an imminent conflict between the neighboring sovereigns, a clash happily avoided, thanks to the spirit of diplomatic conciliation to which the pope added the prestige of his good offices. Brazil came into being as a successful afterthought of the Great Navigations, an epic achievement that brought its distant oriental goal within reach of galleons and caravels by exactly opposite routes—thus confirming the roundness of the earth. It is generally contended that Brazil is a fruit of the Renaissance. More exactly, one should consider it the conscious product of the Renaissance diplomacy that strove to give order and direction to the planetary expansion through seas never previously sailed, owing to the initiative of two Iberian people.

Map 6.1   Tordesillas Line

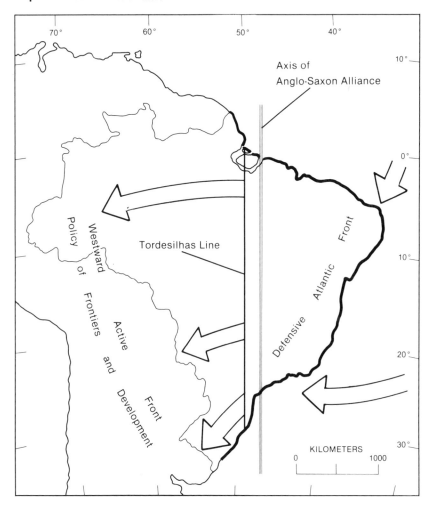

The line of Tordesillas is a geopolitical imposition, the commencement marking the route under which the colonial history of Brazil was arched, thus determining the configuration of our foreign policy ever since. Tordesillas is the first in historical importance and chronological order of the foundations over which Brazilian international life was to develop: it represents Brazil's relations with Spain and with those Spanish dominions to the west of the line that were to become seven neighboring republics. Tordesillas is, therefore, the problem of frontiers, always a question of priority in any diplomatic policy, thus tacitly tabled at the very beginning of discovery by Admiral Pedro Alvares Cabral.

Immediately afterwards, however, a new element in this policy became apparent: the problem of external security along the coast line—from the mouth of the Amazon to the land bulge in the present state of Santa Catarina. The new problem was announced by the ominously ironic remarks of the king of France that he had never heard of the clause in the Testament of Adam that divided the world between his cousins of Portugal and Castille. The *bon mot* of Francis I was intended to legitimize possession of all lands to the east of the meridian of 370 leagues from Cape Verde, it could not achieve a similarly satisfactory result in preventing claims by third parties. These claims came thundering through the guns of French and Dutch adventurers, of Villegaignon and La Ravardiére, of Cavendish, of Jacob Villekens and Hendryk Lonck, of Duguay-Trouin and Duclerc, to cite only a few among the most hostile aggressors. They brought on some hard struggles, attacks by land and tremendous naval combats, considerable efforts, indeed, at coastal defense, shouldered most of the time by the Portuguese settlers themselves and their creole descendants, with very little help from the mother country.

Those European incursions and attempted foreign settlements had, nevertheless, little to expect from the Portuguese diplomatic art, impotent as it was to resist them. Lisbon was involved in the intrigues of the Thirty Years' War and in the struggle between Bourbons and Habsburgs for the control of Europe. It fell for sixty years (between 1580 and 1640) under the rule of Castille, and in its decline it did not even possess resources valuable enough to justify defensive alliances. Thus, only some weapons from the mother country and a few ships flying the banner of Spain in the period of union of the two crowns fought against Dutch invaders by land and sea. The reaction of the local population was the sole strength of the colony during the seventeenth and eighteenth centuries. As early as the seventeenth century, the two fronts over which the policy of security and development of the colony were drawn seemed, therefore, to be clearly delineated: the first front was active, pointing westward beyond the meridian of Tordesillas for expansion into the backlands (*sertão*), attracted by the mysterious unknown and facing as antagonists the Spanish *conquistadores* of the Andean plateau and the Plata area; the second front was essentially passive and defensive, attentive to the seacoast, leaning eastward, and apprehensive at any sudden appearance of hostile fleets that might be looking for new commercial opportunities in the imperial mercantilism favored at the time. This apprehension, incidentally, strengthened the traditional Portuguese tendency toward monopolistic centralization of trade in the hands of the Lusitanian crown. The fronts (see Map 6.1) represent the two dominant preoccupations with development and security required by the fast-expanding colony in its territorial integrity. They constitute, as it were, the two guidelines of colonial foreign policy, two coordinates that invested meaning into the material to be treated by a diplomacy pregnant with possibilities.

Furthermore, because Brazil was discovered as an incident of the great Portuguese navigations, because it long represented merely a stop on the way to India, and because it was colonized with the indispensable work of African slaves, Brazil has a destiny that is not only continental, but eminently maritime—an Atlantic destiny. Geographically, Brazil leans over the South Atlantic. Very early in its history, transoceanic links were established between its Atlantic shore and West Africa, from the Guinea coast and the *Costa de Mina* to Angola. Brazil's Atlantic destiny and communion with the Sudanese and Bantu peoples, as well as its more recent African policy, are perfectly predicted in the intense trade that prospered among Rio de Janeiro, Bahia, and the Gulf of Benin, which was concerned with more than just slaves. They are foreshadowed in the reconquest of Angola from the Dutch in the seventeenth century, by an expedition organized by Brazilians setting out from Rio de Janeiro. The Atlantic role of Brazil is absolutely original in Latin America. Its negative aspect, however, is represented by the stubborn refusal of the population to abandon the seashores and engage in the agricultural occupation of the interior—the process of westward development having begun only during the last fifty years. Yet the double destiny of the country enriches its history and complicates the general outlook of Brazilian foreign policy.

In Europe, Portugal was fighting for its independence against Spain, struggling hard under the leadership of the house of Bragança to ensure that its autonomous development would be respected by the court of Madrid. Lisbon was therefore bound to seek some sort of European support in order to supplement its scarce military resources. During the Thirty Years' War and the subsequent period of French hegemony, the exertions of Louis XIV deflected to the Pyrenees the main thrust of Castillian power. But Spain itself was already in decline. The English alliance soon appeared better suited to the prevailing conditions and therefore lasted longer. It started to take shape in 1661 in a treaty with the restored British king, Charles II, and reached a formal conclusion in 1703, during the War of Spanish Succession. Allied to the Austrians and the English, a Portuguese general entered Madrid in 1706. The alliance between the court of Lisbon and that of St. James would shield the Brazilian coastline, the safety of which was never thereafter to be contested by any other colonial power. The last attempts against Rio were undertaken from 1710 to 1711 by French squadrons under Duguay-Trouin and Duclerc. Thus the security of Brazil was to be preserved from the border wars and territorial exchanges that, during the whole of the eighteenth century, affected the European colonial empires in North America, the Caribbean, and the Orient. The celebrated treaty signed by John and his son Paul Methwen also established the commercial clauses (cloth against wine) of an agreement destined to become a permanent feature of Portuguese foreign policy to this day—a small price to be paid through dependence on the Anglo-Saxon commercial system.

The Treaty of Madrid of 1750 was the second decisive landmark in Brazilian diplomatic history. On this occasion, Alexandre de Gusmão, a Portuguese diplomat and a Brazilian by birth and interest, managed to confirm on behalf of Brazil the new principle of *uti possidetis*. The treaty stipulated that the jurisdiction of the two kingdoms in South America, Portugal and Spain, was to be fixed, taking as reference rivers and mountains, but obeying the actual occupation of territory by their respective nationals. The preamble of the treaty proclaimed that "each party will remain with what it presently possesses." As we may recall, during the time of the union of the two crowns, the Brazilian pioneers (*bandeirantes*) had roamed far and wide beyond the line of Tordesillas, and by so doing, they had practically tripled the colony to its present size of roughly eight and a half million square kilometers. The treaty of 1750 confirmed the conquests of the *bandeirantes*, of the Jesuits, and of the cattle raisers, thus stabilizing, by a legal instrument, the first of the two fronts in Brazilian foreign policy. To the element of action—and even aggression, as exemplified by the assault upon the missions of Paraguay and the foundation of the colony of Sacramento on the left bank of the Plata—that act of high diplomatic foresight added a new principle of collective security and cooperation, foreshadowing feelings of solidarity among all the people of this southern continent.

The Brazilian essayist Alceu do Amoroso Lima has observed that Spanish America came into being by a process of dismemberment and fragmentation of the former Viceroyalties, British North America by the union of the formerly autonomous thirteen colonies, whereas Portuguese America came into its own by a specific process of segregation. The mother country kept the territory in strict political, economic, and cultural isolation, not only in relation to alien European influences, but toward its southern neighbors. The situation prevailed until the decisive turning point in 1808. At that time, the ports of Brazil were opened to international trade by the efforts of the British, who had just saved the Portuguese court from capture by Napoleonic armies. Portugal's insistence on preserving its identity in the Peninsula against the hegemonic ambitions of Castille brought reaction in South America, eventually evolving to Brazil's greater benefit. At the dawn of the nineteenth century, splendid isolation allowed Brazil to watch undisturbed the disintegration of the Spanish colonial empire. The regionalistic and autonomist tendencies of the people of Spain, contained for so long under the inflexible centralizing policies of Castille, finally burst open freely through the immense areas of regional variety of the New World. Brazil remained happily united, faced as it was with the problem of its frontiers. Segregated and clinging together through the ferocious feeling of independence inherited from the Portuguese, the Brazilians maintained without much trouble the unity of their territory. Only two serious revolutionary movements endangered that outlook, one in Rio Grande do Sul and the other in Pernambuco. With all the tremendous potentialities that such circumstances grant the country, union in the immensity of its territory is the sole miraculous datum in the history of Brazil.

A SCHEME FOR BRAZILIAN GEOPOLITICS

Three principles have been proposed as leading Brazilian foreign policy for the attainment of its permanent national objectives:

1. The preservation of the frontier line against the territorial ambitions of neighboring Spanish-speaking republics: this is called the "Frontier policy."

2. The defense of Brazilian territorial supremacy in South America against any attempt at the reconstitution of the old Spanish viceroyalties, particularly in the La Plata region: this is called the "policy of Equilibrium." It justified the hostility of Brazil toward Rosas of Argentina; Solano López of Paraguay; and, more recently, Perón, in his wild imperial dreams.

3. Closely connected to the preceding principles is the protection of the internal political stability of the country against the spirit of *caudillismo*. From this resulted the "policy of Intervention" as happened in Uruguay, Argentina, and Paraguay in the nineteenth century.

The first two principles, as can be seen, have today a mere historical significance. They represent steps that have already been transcended in the search for permanent national objectives. The third principle, however, entails the preservation of internal order through operations aiming at avoidance of contamination by anarchy, despotism, and totalitarianism. As a result of the impact upon Latin America of extracontinental ideologies inimical to the democratic way of life, this policy is by no means obsolete. A modern version justified Brazilian participation in World War II, when an expeditionary force was sent to Italy (1944 to 1945). The principle of intervention also accounted for Brazil's presence in the occupation of the Dominican Republic (1965), when that country was endangered by a communist-oriented military coup; and the breaking off of relations with Cuba (1964 to 1986). Nowadays, this policy does not find the same acceptance, for reasons which will be seen shortly. It has been corroded by anti-U.S. sentiments like those found in the Third World, seen, for example, in the case of Nicaragua, where Brazilian diplomacy has tended to favor the Sandinista cause.

Other scholars who have studied Brazilian history tend to emphasize the problem of dealing with Argentina in the establishment of Brazil's land frontier in the south as the most strenuous Brazilian diplomats have had to face. In fact, this problem has played a formative role in Brazilian diplomacy, as far as long experience, solid institutionalization, and training of specialized professionals are concerned.

Another point is worth raising at this stage: expansion of the human frontier due to demographic growth, the building of roads and railways, and the extension of airlines, facing a contrary movement eastward from the side of Bolivia and Peru, where the population is beginning to drive down from the Andean altiplano to occupy the Amazonian jungle. This double movement will soon clash, creating living frontiers in confined areas that were, until recently, purely abstract geographical lines. This transition occurs also from the side of Venezuela, a country that is undergoing a rapid growth as a result of the development of its mineral wealth. The building of Brasília has contributed to this result. In Paraguay, for instance, half a million Brazilian farmers and the joint building of the Itaipú hydroelectric power project will certainly have a decisive impact upon the destiny of that country cut off from the seas.

In the evolution of Brazilian foreign policy, therefore, we can distinguish three successive historical phases: in the first one, the human frontier of the colonial nationality expands toward the interior, beyond the Tordesillas meridian, facing the Spanish settlers in the distant reaches of the Guaíra and Sacramento colony, in the extreme south; in the second phase, Brazilian diplomacy tries to consolidate and institutionalize the borders thus conquered—this is the phase that culminates in the work of peaceful negotiations and arbitration under the administration of the famous foreign minister Baron Rio Branco; and in the third phase, up to the present, the dynamism of diplomatic activity is converted into an effort at peaceful cooperation with Brazil's newly rediscovered Hispanic neighbors. Simultaneously, Brazil becomes conscious that it has an extra-continental destiny to pursue, especially in Africa. In other words, Brazil is beginning to come of age.

American independence, the French Revolution, and the Napoleonic wars started a process at the end of which the whole of Latin America, Brazil included, obtained self-determination. Important diplomatic intrigues took place and decisive foreign policies were followed after those episodes. There were timid attempts from the Brazilian side to obtain the interest of, and court favor from, the new leaders of the former British colonies in North America on behalf of Brazil's own whims of independence. These were considered, however, too imaginative and deprived of realism. Yet certain scholars have considered those inchoate attempts in Washington as the foundation of the bonds of friendship between Brazil and the United States. They were unilateral steps to be sure, and the American reaction was one of indifference. But, whatever it was, the pursuit of the U.S. alliance could be naturally deduced from the role played by Great Britain to the benefit of the integrity of the Portuguese-speaking territories in the New World. Emperor Pedro II pursued that enlightened policy. In 1876, he travelled to the United States for the centennial, the first South American head of state to do so. The traditional axis of Anglo-Portuguese alliance was simply transferred to the Americas, and for the same reasons. The idea took some time to mature—a hundred and fifty years, as a matter of fact—but the seeds of the grand scheme were sown very early, as an intuition of a geopolitical necessity.

Our attention is also drawn, as far as this question is concerned, to the Monroe Doctrine, which made very little sense or had a limited value for Brazil, at the time it was promulgated. Whatever the historical value of the doctrine in the elaboration of the Pan-American idea, it was not meant for Brazil: Portugal in the 1820s and 1830s had no power or material means to cherish any hope of reconquering Brazil. Besides, the monarchical system Brazil had adopted under a crown of legitimate European lineage kept the country well protected from the wrath of the Holy Alliance. A few years after independence, Brazil already belonged to the family of civilized nations, surrounded, perhaps, by the mysteries of tropical exoticism, but enjoying a sort of respect that, at that time and outside of Europe, only the United States itself had managed to acquire. A Venezuelan president mused sadly, when Dom Pedro II was exiled in 1889, that Brazil and Chile were the only two organized democratic countries in the hemisphere, "the republic" of Brazil and the "empire" of Chile.

Because of the prestige of the Empire and the internal order and stability it enjoyed—in contrast to the anarchy then prevailing in most of Latin America—Brazil preserved its self-determination without being exposed to any particular risk. Brazil even engaged in several wars in the La Plata region without provoking undesirable foreign interventions. If British commercial interests contributed to pay for a discreet naval protection, it certainly was not an exorbitant price. Brazil enjoyed a vast unexplored territory (certainly aplenty with unknown riches) and an immense, unprotected Atlantic coastline, open to all temptations. The risks that were avoided were certainly considerable if we take into account what happened in the late nineteenth century to most of Africa and Asia. On only one occasion foreign intervention loomed possible: in 1893, soon after the republic was proclaimed, a short civil war pitted army radicals against navy conservatives in the Bay of Rio de Janeiro. Foreign warships were caught in the cross fire and warned the government that they might disembark marines. They refrained from doing so when President Floriano Peixoto replied that foreign troops would be received with bullets.

The removal from London to Washington of the third axis of Brazilian foreign policy took definite shape only at the beginning of this century, under the enlightened administration of Baron Rio Branco who both consolidated and administered this policy. "We find ourselves at the beginning of a new era," wrote Joaquim Nabuco, a prominent Republican statesman who was appointed first full ambassador to the U.S. capital. "In our calculations, the observation post in Washington is the most important. . . . Under these conditions our diplomacy should be made principally in Washington."

In other words, in the Brazilian conception that developed at the beginning of this century, the American continent is divided into three, not two, parts: Anglo-Saxon North America, Spanish America, and Brazil—Brazil playing precisely the role of a "third force," burdened with the task as unofficial go-between, interpreting and settling disagreements in the great dialogue of the Pan-American community. This "third position" is still Brazil's. Nevertheless, we should emphasize that it has never been well understood by the Spanish-Americans, who consider Brazil under a sort of vassalage with regard to the United States; nor by the "Anglos" who never clearly distinguished Brazil—one among twenty-odd Hispanic republics—from the other "Latinos." Most commonly, U.S. citizens don't even know that Brazilians speak Portuguese, not Spanish. In the interplay of these ponderable historical-cultural, psycho-social, and geopolitical factors, the inter-American system was created.

We can ascertain, in conclusion, that by the time Rio Branco took over the Ministry of External Relations, the North American alliance loomed large as one of the foundations of Brazilian foreign policy. Positive and negative aspects, offensive and defensive, dynamic or passive, could also be ascertained in relation to Brazil's Spanish-speaking neighbors, particularly those of the La Plata region. Brazil used to feel surrounded, isolated in a continent of Hispanic nations. There was even a certain amount of snobbery in its attitude toward them. But Brazilian reaction reached beyond the effects of the inferiority/superiority complex, the same way it overstepped the narrow limits of Tordesillas. The actual frontier, lost somewhere in the Amazonian jungle, or *Mato-Grosso* (the "thick bush country"), the frontier of struggle, of advance and retreat, of intervention or pioneering incursions, the living frontier of commerce, cooperation, and mutual involvement along the Uruguay and Paraguay rivers all have become frontiers of creative tension where Brazil can already devise the ideal image of a future and necessary political integration.

A FOREIGN POLICY PERSONA

Early tendencies of its policy toward the world found expression in what I call the Brazilian *persona*. The new nation endeavored to strengthen its national security by adapting its governmental system to what seemed more in vogue in Europe. By becoming an empire ruled by a European royal family, it guaranteed its independence against a possible colonial return of the reactionary forces of the Holy Alliance. By turning into a federal republic according to the U.S. model at the turn of the century, it again guarded itself against the rising tide of European colonial imperialism. Between two World Wars, the country toyed with fascist political schemes then in the upswing. Vargas inaugurated the game of playing one dominant power system against the other. During a short period (1937–1940), he applauded Hitler's New Order, as witnessed by a famous speech in June 1940; he turned to left-wing support for his staying in power in 1945, and for his return to power from 1950 to 1954.

Brazilians always displayed a widespread preoccupation with the manner in which Brazil is perceived by North Americans and Europeans. There is an impatient ambivalence in the way in which they wish to be considered and judged. I maintain that this attitude reveals a certain inferiority complex; this results from Brazil's having made North America into its "model society." In political terms, this ambivalence reflects itself in the argument whether Brazil is or is not a Third World country, or a friend and independent competitor of the United States. On the other hand, this provokes a certain inconsistency in intellectuals: they sometimes describe themselves as ill-treated supplicants who deserve the right to receive more from their wealthy and egocentric persecutors from the North; at other times, they pretend to be independent and equal coactors, feeling consequently offended by classification as "Third Worldly" or dependent on multilateral acts of charity.

We could focus on two points which have to do with the Brazilian *persona*. First, many Americans (not necessarily for reasons of flattery) have insisted on the similarities between Brazil and the United States. Like the United States, Brazil is a vast, multiracial country, possessing many natural resources, a country that liberated itself from a European colonial power. Brazil is populated mainly by European immigrants. At the end of the discovery, the Europeans encountered indigenous peoples who were "culturally inferior." This immediately distinguishes Brazil from Mexico and from the Andean nations, and puts the country on a different platform. São Paulo is indeed much more like a North American metropolis than any Hispanic city, including Mexico City. Brazil has known African slavery, which, however, was abolished in a milder way than in the North American South. Brazil prides itself on "a conquest of the Far West" of epic proportions, the growth of which is still taking place. It has built a new capital city and is a culturally well-decentralized federation. Other features common to both the United States and Brazil are the gold rush, industrial expansion, and the opening up of new pioneering territories (albeit with 50 to 100 years' difference in time). Brazil's Catholic/Latin religious and cultural background, however, situated in the zone of the humid tropics, gave rise to radical socio-political differences between it and the United States. The second point is that, in common with its friends from the North before World War II, Brazil has nurtured a secret penchant for isolationism, with a nearly autistic consciousness of its own reality. Under such conditions, it is not surprising that there is an equivalent reaction to the warped U.S. perception; this has provoked the ignorance, indifference, and even the sense of superiority Brazil has in relation to its continental neighbors—with the exception of Argentina, where a mutual compromise has been brought about by the dictates of history. Who, in Brazil, would be able to distinguish clearly Ecuador from Colombia, or Costa Rica from Honduras? Brazilians have no right to complain about the U.S. attitude, because their own also contains a large dose of *snobismo* regarding their neighbors and Hispanic cousins.

The deepest contrast between North and South America is political. It rests on U.S. knowledge about self-government and mastery of political economy, both of which Brazil lacks. That is why the "gringos" are rich and powerful, while Brazil is underdeveloped. They succeeded in establishing firm and durable bases for a system that reconciles order, justice, and freedom, defined in terms of democracy. Brazilians have never overcome the antinomy and remain under a regime that could better be termed "patrimonialist" (in Max Weber's sense of the word). The result is that Brazil oscillates between periods of anarchy and periods of overbearing authoritarianism. Since it has shown no talent in the art of self-government, it has failed to gain the respect of North America. It is, of course, not up to the United States to take this part of the world seriously. This will happen only when countries like Brazil decide to take *themselves* seriously and behave accordingly. It shall cease to be an "invisible continent" on the day that the Brazilian *homo ludens* transforms himself into *homo sapiens* and *homo faber* . . .

chapter seven

Traditional Themes of Brazilian Geopolitics and the Future of Geopolitics in the Southern Cone

Philip Kelly

Among important elements comprising the geopolitical aspect of foreign affairs, three in particular deserve special mention when describing traditional and contemporary themes of Brazilian geopolitics. First is *distance* from Eurasia and North America, which, in the case of Brazil, implies security, isolation, independence, and policy flexibility and continuity. The spatial element of *position* in the core of South America brings dichotomies of a Brazilian center and a Spanish periphery, of pivotal location and exclusiveness, of maneuverability and containment. Third is a natural *resources* base, reflecting in Brazil expansion of regional and global impact, as well as dependency and vulnerability. I see a mixture of these three elements contributing to a Brazilian foreign policy of moderation, balance, and confinement in its relationships toward Latin America and toward other world continents. But, simultaneously, this complexity encourages a policy of boldness and leadership in the Southern Cone as well.

At least six rather distinct but interrelated traditional themes of Brazilian foreign and security policy can be gleaned from the geopolitical elements above:

1. A Brazilian position located on the global periphery, relatively distant from Eurasian or northern struggles and submerged as a large power in the less strategic southern world
2. A central continental position that faces encirclement by wary neighbors in addition to opportunities for expanding territory and for enhancing Southern Cone integration with neighbors
3. A persistent rivalry with Argentina that has created tensions in the Southern Cone, retarded integration, and blocked both Brazil and Argentina from exerting a more influential global role
4. A natural resources and population base sufficient for continental leadership and for greater world status

5. A national potential for social, political, and geographic fragmentation
6. A geopolitically oriented foreign policy, in large part derived from the distance, positional, and resources elements, as well as from the five themes described above

Part One of this chapter examines each of these six themes within the tradition of Brazilian geopolitics. Next, in Part Two, I will briefly project these traditions into a possible Southern Cone strategic configuration, positing that a heightened world significance for the region could arise, based upon a Brazilian-Argentine rapprochement that could transform events in South America and beyond.

TRADITIONAL THEMES OF BRAZILIAN GEOPOLITICS

In this chapter, geopolitical "themes" signify broader categories or dimensions containing interrelated linkages in foreign affairs. These descriptions all relate to the Brazilian world location and to its leaders' perceptions of this space in relation to other political spatial attributes. My aim is to point out indicators within a geopolitical framework for interpreting Brazilian foreign policy, then expanding these into an integrationist scenario which may now be emerging.

Theme One: A Brazilian Position Located on the Global Political Periphery

An essential feature of the global thrust of twentieth-century geopolitics is a focus upon Eurasia, that largest of land masses, spanning territorial spaces from Europe through Asia and containing roughly two-thirds of global population and resources. Occupying a southern and peripheral position, Brazil is largely excluded from this attachment. The Eurasia thesis predicts likely confrontation among northern Great Powers for domination, or to prevent domination, of certain vital areas on or around this land mass, and, in most cases, dominance of Eurasia translates into global hegemony. Another geopolitical concept of global scope, the dependency thesis, pits northern capitalist nations against the Third World southern periphery, including Brazil, although in this situation, North Atlantic attempts of exploitation are contested by the South. In both of these and most other geostrategic models, only northern environs are those "which count."

Brazil's isolation from Eurasia's power struggles and, consequently, its strategic independence, I contend, are nevertheless as much to Brazil's advantage as to its disadvantage. It does not indicate lack of potential for global impact, nor even preclude some type of strategic hegemony, for Brazil possesses resources itself and access to those of neighboring southern territories and marine spaces. In geopolitics, we think in terms of strategic linkages, and Brazil clearly has the means and geographic position for taking advantage of northern strategic stalemate and conflict to assert a world status. Brazil's growing influence in contemporary world affairs and role as a leading Third World nation already attest to this potential.

In this spatial nexus, Brazil resembles the peripheral positions of Australia and South Africa, all three nations being allied to (but not dependencies of) Western democracies, having access to abundant natural resources, enjoying territorial exclusion from direct involvement in Eurasian/North American military confrontations, and possessing important strategic maritime choke points. The difference between Brazil and these other two states—and, in my judgment, the key to her foreign participation—rests in the possibility of Brazilian attainment of Great Power status and in its continental encirclement by potentially hostile neighbors who can prevent such status.

Brazil likewise approximates the continental position of the United States sans Brazilian occupation of the Pacific coast. Both republics dominate their respective halves of the hemisphere, both possess interior river systems and ample resources, and both contain heterogeneous populations. Until North American adoption of cold-war internationalism, both exhibited similar foreign and security policies of isolation and foreign trade. Obviously, the two countries differ in other important ways: the northern location of the United States within a more direct Eurasian dimension of encirclement, its greater natural wealth, and its relatively weak continental neighbors.

But Brazil possesses certain areas of strategic value to the North: (1) proximity to several important maritime choke points: the mouth of the Amazon, the southern passages of South America, the Atlantic Narrows, and the Cape of Good Hope, these latter places being vital primarily to petroleum traffic destined for the North Atlantic; (2) possession of the Brazilian "Bulge," the closest Atlantic access between America and Eurasia; (3) abundant wealth and population, improving technology, and the development of its hinterlands; and (4) a pivotal position for stabilizing and integrating the South American nations. It is my assumption that these features of strategic power will increase in geopolitical value commensurate to the continuation of East-West tensions in Eurasia and to the weakening influence of the United States and the Soviet Union in world affairs.

Ties between Brazil and the United States traditionally have been cordial, although, for the most part, permanent bonds have not arisen because of North American isolationism, its failure to visualize Brazil's strategic potential, and its preoccupation with Eurasian relationships. Despite neglect by the United States in strengthening a hemispheric tie, a *barganha leal*, or deal, has normally been Brazil's stance toward North America, Brazil gaining status and economic assistance from its disinterested northern benefactor, the United States satisfying its security and investment objectives.

From a geopolitical standpoint, it is surprising that the United States does not forge closer strategic linkages with Brazil. The two nations are natural allies in the sense that their spheres of influence do not overlap, they share similar foreign policy traditions (neither condone extracontinental intrusions into America that might destabilize frontiers), they occupy continental locations peripheral to Eurasia, and their diplomatic heritages are American, Western, pacific, and commercial. What may be impeding a strategic Brazil–United States linkage is the U.S. perception that Brazil lacks a power capability and hence cannot contribute decisively to American defense, a calculation probably true at the present moment.

Theme Two: A Central Continental Position

Westward expansion to the South American core and beyond represents a predominant feature of Brazil's traditional geopolitics. One could possibly have predicted that vigorous Portuguese settlers, having consolidated coastal enclaves, would push inward toward the continent's core, utilizing river and land passages to stake out claims in the vacant hinterland. To the present day, we see an organic extension along distant frontiers, particularly those of Paraguay, Guyana, Bolivia, and Uruguay. This advance resembles the expansion of the British in North America, both peoples finding themselves in positions of continental pivot and, consequently, of military, diplomatic, and commercial advantage.

Four geopolitical doctrines in particular describe Brazilian expansion: the doctrine of organic frontiers, the South American heartland thesis, the concept of Brazilian Manifest Destiny, and the law of valuable areas (Kelly 1987, 80–83). The organic theory of the state depicts a cycle in which states instinctively compete for space, into which the victorious broaden their power, and the vanquished contract in territory. Brazil's geopolitical writers (Meira Mattos 1977; Golbery do Couto e Silva 1967) favored this conception because it provided a rationale for previous expansion as well as a format for national development and leadership in regional integration.

Brazilian strategists have also been drawn to the South American heartland concepts of Travassos' (1947) "magic triangle" and Tambs' (1965) Charcas triangle (in both cases, the strategic space between the Bolivian cities of Sucre, Cochabamba, and Santa Cruz de la Sierra), and of Golbery's (1967) "continental welding zone" (encompassing Bolivia, Paraguay, and south central Brazil). Because these locations control the headwaters of the Amazon and of the La Plata watersheds and also major passes over the Andes to the Pacific, some allege that dominating these pivotal zones could give Brazil mastery of South America.

A Manifest Destiny theme assumes eventual Brazilian possession of Pacific coastal lands and islands. Reflecting Spanish American fear of this penetration, a Peruvian official remarked: "Brazil, which is like the United States a hundred years ago, believes she has a manifest destiny to occupy the continent and reach the Pacific, and Peru is California" (Child 1985, 98). In all directions, Africa and South Atlantic in addition to this hemisphere, Brazil's impact appears to be expanding, prompting its geopoliticians to theorize a natural propensity for spatial growth.

Several scholars (Meira Mattos 1977; Travassos 1947; Pittman 1984; Tambs 1966) typify historical Brazilian expansion according to a progressive settlement of valuable but vacant interior territories: first, private Brazilian exploitation of isolated lands of uncertain national ownership, then eventual state protection and sponsorship, and, finally, arbitration and annexation to coastal ecumenes. Contemporary cross-border settlements in Uruguay, Paraguay, and Bolivia may resemble this organic tendency, spawning fears of loss of an Amazonian "buffer" protecting their vulnerable interior possessions (Bond,1981, 128).

Brazil's central continental position, historical expansion, and growing power during the twentieth century contributed to a perception of encirclement, Brazil being a "Portuguese American island in a Spanish American archipelago . . . surrounded by latently hostile . . . neighbors" (Hilton 1975:14). Yet Brazilian security never was seriously threatened by this enclosure, potential adversaries being either too weak, divided, or isolated to impede Brazil's territorial ambitions, although one could conclude that such encirclement probably dampened the most extreme Brazilian hegemony during the earlier period.

To its neighbors, Brazil responded with policies of isolationism and aloofness, or of bilateral diplomacy, or of participation within a balance of power configuration natural to the Southern Cone. In this diplomacy, Brazil sought to avoid destabilization of its frontiers, to prevent Argentina from assembling American allies against Brazil, to maintain the independence of the buffer states (Bolivia, Uruguay, and Paraguay), and to project influence overseas in quest of commercial profit and attainment of *grandeza*, or world-power recognition. These approaches, in my opinion, reflect the positional, spatial, and resources dimensions of Brazil's traditional place within South America.

Theme Three: A Persistent Rivalry with Argentina

Although neither state has directly confronted the other in war, the two have remained wary of each other in matters of security, commerce, and continental leadership. . . during much of their national existence. Frank McCann concludes, "What is known is that Argentina and Brazil are each other's most probable opponent in their respective war plans" (1983, 299). The major reasons for this confrontation appear to be two: the quest by both states for Southern Cone primacy—the Argentineans now blocked by Brazilians, who clearly lead (Selcher 1985, 27) in continental power statistics—and competition for control of the buffer states of Bolivia, Paraguay, and Uruguay, the victor in this struggle apparently the emergent hegemon of the region.

One reflection of the Brazilian-Argentine rivalry is a South American balance-of-power configuration that aligns Brazil with Chile, Ecuador, and Guyana against Argentina, Peru, and Venezuela (Child 1985, 175–179). Within Brazilian diplomacy, this balancing phenomenon aimed to prevent Spanish encirclement, to maintain the independence of the buffer states as well as of Chile, and to allow pursuit of major power status in overseas affairs. Nevertheless, it likewise created a potential for escalation of serious Southern Cone conflict, particularly with Argentina, in the buffer states, and of conventional or nuclear arms races.

Brazilian-Argentine rivalry has hindered Southern Cone integration, created possibilities for serious, indigenous nuclear weapons development, jeopardized peaceful settlement of disputes in the region, intensified competition among Southern Cone states for control of the Antarctic, and prevented Brazil from distancing itself from the United States. Above all, the cleavage between Brazil and Argentina has forestalled a more assertive role for the region in world strategic relationships, a checkmating effect that has kept the entire area fixed to the global political and economic periphery.

Theme Four: A Significant Natural Resources and Population Base

Since colonial times, Brazil has pursued *grandeza*, a quest for world power recognition. By the mid-twentieth century, this status was well along the way to attainment, after impressive national gains in industry, trade, technology, and resources exploitation (Selcher 1985, 27; Meira Mattos 1977). Such recognition drew further stimulus from U.S. endorsement and from the stagnation of Argentina, Brazil's traditional South American rival. General Carlos de Meira Mattos exemplifies this euphoric spirit by stating (1977, 143):

> We will be a world power, reaching this goal in our development by [the year] 2000, and this will be independent from our vocation or enjoyment of such power. We must be, therefore, prepared also to exercise this power, protecting our interests whose dimensions in terms of economics and geostrategy will acquire world amplitude.

This prestige, traditionally assumed by Brazilians to be their special destiny, translates into economic development, national integration, and frontier security, but seldom has it corresponded to military projection within and beyond Latin America (Kelly 1987; Meira Mattos 1977; Schneider 1976).

Although its leaders deny a quest for regional hegemony, Brazil's recent rise to South American leadership has substantially altered the geopolitical framework of continental power relationships, in these ways: it has shielded South America from North American interference, cold-war strife, and the entry of foreign nuclear weapons (although the nuclear void could be quickly filled by local production in either Brazil or Argentina); it has made Brazil the essential continental pivot in terms of regional security and political stability, of economic development and prosperity, of coastal-to-interior and transnational integration, and of extracontinental bargaining.

It would be an exaggeration, nevertheless, to argue that Brazil, within its own means, could alone rise to Great Power status within the global political arena. Although it possesses resources, Argentina and other Spanish states, despite their being weaker than Brazil, checkmate Brazil's ability to project effective power overseas. Only through Southern Cone integration, in my opinion, can Brazil break through this continental confinement and play a geostrategic role in global affairs, but in this case, not as a national actor but as a leader of a larger regional, that is, Southern Cone, coalition.

Theme Five: A National Potential for Social, Political, and Geographic Fragmentation

Throughout its history, Brazil has contended with national fragmentation. The Atlantic coastal fringe was first settled by the Portuguese, and the focus then remained largely on this littoral, its center eventually shifting from Salvador, Bahia, southward to the triangle of Rio de Janeiro, São Paulo, and Minas Gerais, where today are concentrated the country's dominant cultural, industrial, technological, governmental, and population ecumenes. Consequently, national territory beyond the coastal strip never became fully developed or integrated with coastal regions, and distant hinterlands as well as rival coastal centers continued to be disintegrated and exposed to potential foreign exploitation or regional autonomy (Sternberg 1987). Would the Amazon someday fall under the control of another country, or itself become an independent region?

Two additional features in this centrifugal rhythm also require brief mention. Western expansion was brought about through diplomatic and/or military pressure, not by economic colonization, and territorial growth stopped short of acquisition of Pacific coastal zones, largely because of distance, of Spanish settlements in the Andes and the La Plata Basin, and of impenetrable Amazon jungles and Andes mountains. Hence, less economic incentive nowadays exists for interior consolidation, and such areas remain largely dormant as a result. Furthermore, all national population clusters contain wide racial, nationality, and class divisions. Unification along these lines could not transpire because serious wealth gaps and social marginality continued, despite industrial growth, and the Amazon frontier has not proven to be, for immigrants, a racial or class equalizer.

National unity, particularly in a geopolitical sense, is vital to Brazil's future. General Meira Mattos, for example, raises the possibility of rebellion in the Amazon: "The enemy is now within [Brazil], not a threat of direct attack across our borders . . . .The real international threat is [internal] revolutionary war" (1973, 63). This internal-security theme likewise is emphasized by General Golbery de Couto e Silva (1967). The Amazon's sparse population and untapped natural wealth may also attract attempts by foreigners toward acquisition (Meira Mattos 1980, 93):

> It would be dangerous to leave the vast Amazon basin empty and underdeveloped, when there are areas experiencing grave overpopulation—Bangladesh, Indochina, and Japan. [It would not be] desirable to the Amazonian countries to lose their sovereignty over this coveted region under the pretext of their incapacity to exploit it.

Utilization of jungle wealth is assumed an important requisite to development and to *grandeza*, linking national integration inexorably to security requirements.

In tones of Toynbeean "challenge and response" toward confronting the Amazon hinterlands, Meira Mattos declares that consolidation of Brazilian regional domains will create a new "tropical civilization" headed for South American leadership (1984), and this new spirit will propel Brazil toward overcoming:

> the problem of conquering its own territory, of integrating its maritime and continental portions, revealing to us the importance of a strategy of territorial transportation and population, the capacity of materializing [projects] within the context of geography, and a political will of state (1980, 166–167).

A variety of official methods have sought connection of interior regions with coastal ecumenes: a federal and frequently authoritarian structure of government; the construction of an interior national capital at Brasília; the erection of interior highways, settlements, and communication centers; the promotion of joint development projects with neighboring states; the stabilization of frontier strife by means of military force and assistance (particularly in the cases of Paraguay, Guyana, and Suriname); the maintenance of low-keyed diplomatic and military policies; and alignment with the United States and the Western alliance.

Theme Six: A Geopolitically Oriented Foreign Policy

Despite claims of imperialism and manifest destiny by critics, Brazilian foreign policy typically has been cautious, moderate, and capitalist in nature through much of the twentieth century (Selcher 1986; Kelly 1987). This perspective is in large part, I believe, reflective of the country's geopolitical heritage, born of a continental stance that requires careful balancing of advantages and liabilities relative to its location and resources. Brazil's geographic position, peripheral to northern land masses and balances of power, yet within the core of South America, and its ample natural resources, demographic concentration along the South American littoral, propensity for fragmentation, and industrial and technological potential, have determined, in my opinion, a fairly clear and consensual orientation for consistent international conduct.

Consequently, Brazilian diplomats balance the advantages of pivotal position with the dangers of Spanish encirclement and of internal divisiveness. If it decides to expand toward the Andes and La Plata, it faces internal disunity and external attack. Brazil finds itself in a convenient position for alliance with the United States; nevertheless, too-cordial northern ties hamper its flexibility in southern-world diplomacy. Although its power alone is not sufficient for global leadership, stalemate and conflict in the North could increase Brazilian influence beyond what could be expected from the country's natural resources. Commercial and security goals could be furthered by aggressive policies toward both northern and southern worlds, but these policies also require, in my opinion, harmony and integration in South America.

A foreign policy consensus along geopolitical lines has emerged, in part, because of successes in historic expansion beyond original coastal settlements and because of the contributions by geopolitical theorists to decisionmaking in security and development matters. Brazilians have been fascinated to this day with the South American continental core they could eventually possess or come to dominate. It appears that Brazil was probably in the best position, in contrast to its neighbors, for expansion toward the interior. Seven major treaties, from that of Tordesillas in 1494 to those with Venezuela (1905) and Colombia (1907) consummated a thirst for acquisition. With the exception of the concluding agreement after the War of the Triple Alliance against Paraguay (1872), all territorial gains were attained through diplomacy. These successes contributed to the respect and momentum given the traditional impetus in Brazilian foreign policy.

Prominent geopolitical writers (Backheuser, Delgado de Carvalho, Travassos, Cartaxo, Rodriguez, Graça, Golbery, and Meira Mattos) supported policies of territorial expansion and national integration (Child 1979, 93–94), and several urged further projection of Brazilian interests to the South Atlantic and the Antarctic as well. In fact, one is able to see a continuum of most geopolitical ideas and policy recommendations spanning the past one hundred years, each theorist accepting and expanding themes of predecessors. As derived from these outstanding and insightful theorists, this geopolitical consensus provides a depth and a consistency to foreign policy not experienced by other Latin American nations. In addition, several Brazilian geopoliticians, particularly the Baron Rio Branco, Golbery do Couto e Silva, and Meira Mattos, held positions of political authority that enabled their ideas to be applied to domestic and foreign policies, adding to the importance of the geopolitical orientation of Brazilian foreign affairs.

TRADITIONAL GEOPOLITICAL TRENDS PROJECTED TO THE FUTURE IN THE SOUTHERN CONE

It seems to me that the key variable linking traditional themes of Brazilian geopolitics to future foreign policy projections is whether the current rapprochement between Brazil and Argentina will persist into the next century, and even strengthen to an entente status. This contemporary possibility of common strategies between the two nations could represent a major watershed in the geopolitics of South America and in the continent's ties to North America and to world politics in general. Likewise, it provides Brazil an opportunity to break her traditional geopolitical shackles and to construct a more flexible foreign policy different from previous stances of caution and balance.

Whether the present rapprochement is a temporary or a permanent condition is difficult to predict. But several recent policy shifts indicate that a harmonious Brazilian relationship with its Southern Cone neighbors will be durable. A consensus apparently has appeared in Brazil during the past decade that "its national development could not be planned or achieved separate from the Latin American context" (Selcher 1986, 68–69). As a result, Brazil has resolved disputes with Argentina and Paraguay (primarily the Itaipú-Corpus controversy) and promoted integrative projects for the Amazon, for hydroelectricity, and for intraregional trade. The major thrust of Brazilian diplomacy has turned to constructing amicable relationships with neighbors, and this to date has brought the desired harmony (Bond 1981). Brazil has loosened ties with the United States to the point of no longer being viewed in South America as a North American proxy, a transition adding to closer regional ties. There also appears to be more interest on the part of Argentina and the other countries toward pursuing the benefits of cooperation with Brazil. The new ideal seems to be that economic integration and peaceful settlement of regional disputes enhances the welfare of all countries concerned and that such a momentum of progress should be extended (Quagliotti de Bellis 1986).

In contrast, likely issues contributing to a disruption of Southern Cone integration could be the reemergence of a Brazilian-Argentine military contest, including indigenous nuclear arsenals; rivalry over buffer states, particularly if any of these succumb to instability; rise of belligerent governments in either Brazil or Argentina and a break in rapprochement as a consequence; disruptive U.S. involvement in South American affairs; and a shift in Brazilian or Argentine economic and political attention away from South America and toward participation in other world concerns. Such factors, however, may not arise to the extent of disturbing momentum toward regional interactiveness, because of the independence of South America from northern interference, the growing realization of benefits gained from integration, and the peaceful overlapping of Argentine and Brazilian foreign-policy objectives. Assuming the rapprochement between Brazil and Argentina will persist, I would like to draw, in the few paragraphs remaining, several possible geopolitical configurations in this transformed Southern Cone relationship.

Under these new circumstances, emergence of a nuclear arms race between Brazil and Argentina would be most unlikely. Although both possess technology and resources for building such weapons, neither would gain an advantage by doing so. I agree on this matter with Steven Gorman, when he writes (1979, 63): "Brazil would have nothing to gain and virtually everything to lose in shifting from the present conventional balance of forces (which favors Brazil) to a strategic balance (which might favor Argentina). . . . Nuclear weapons would serve to heighten the destructiveness of warfare [in the region] while probably not changing its outcome." Nuclear weapons in the Southern Cone obviously would terminate the Brazilian-Argentine rapprochement, but their introduction also could attract U.S. and Soviet strategic interest, and bring the escalation of a shatterbelt condition to the area. Certainly, regional tensions would broaden as a result, and integration and diplomatic maneuverability would seriously atrophy. Nevertheless, I again argue that present trends do not appear to substantiate this nuclear scenario; in fact, most signs indicate a determination among leaders for closer, rather than distant, intraregional relations and a continued preference for excluding northern influence from the area.

A Brazilian-Argentine rapprochement and, better, an eventual entente could bring widespread dividends to Brazil and to the Southern Cone. For one, economic integration, not possible during periods of regional friction, could blossom and prosperity emerge, with the construction and linking of transcontinental road and rail systems; the coordination of industrial, energy, technological, and natural resources; the extension of trade and travel; the development of wealth in the hinterlands, sea waters, and Antarctica; and the economic advancement and integration of Bolivia and Paraguay with the coastal nations. Closer integration of security relationships might bring continued progress in peaceful settlement of regional disputes (for example, Bolivia's quest for a Pacific outlet, the overlapping Antarctic claims, and Ecuador's demand for the Marañón territory); common political and military strategies toward eventual control of the Malvinas, the Antarctic, and seabed resources, bolstered by stronger and coordinated naval forces; the removal of the persistent stigma of Brazil's performing as a U.S. proxy; and a united front against northern manipulation of southern affairs and for the advancement of Southern Cone interests.

To think of Brazil alone as a probable superpower or as Great Power broker is quite unrealistic. By itself, it lacks resources for elevation to world significance among the great northern powers. It global position is peripheral and not of major strategic importance to North America or Eurasian competitors; it is not a continental nation, one which extends from Atlantic to Pacific or Caribbean, and this expansion likely will not occur. Rather, Brazil contends with internal fragmentation and with Spanish encirclement, particularly a flanking by Argentina, its persistent rival. Brazil's place is, instead, clearly the Southern Cone, and, I would argue, its goals must aspire to entente with Argentina first, then to regional integration and peace, followed by a role of leader of Southern Cone autonomy and unity against northern intrusions. Eventually, this common front could bring equal Southern Cone partnership with the United States in reducing the scourge of nuclear war and in maintaining a Eurasian balance of power. To my mind, these are Brazil's geopolitical contributions to the world, the country's destiny to be a member nation of the Southern Cone but, at the same time an important regional participant in the aspiration of global peace.

part four

THE LA PLATA BASIN AND BOLIVIA

chapter eight

The La Plata Basin in the Geopolitics of the Southern Cone

Bernardo Quagliotti de Bellis

Geographic spaces are defined in historical time and become dynamic as a function of economics. Geographic spaces are very complex and unstable entities, while historical times are not produced spontaneously, and economic issues are managed by means of specific laws. We must also consider social and political events, which are the product of particular conditions subordinated to the will of man.

The starting point of any geopolitical analysis is critical realism, an attitude requiring the study of physical facts in a determined space, along with the life and history that have been evolving in it. This brings us to consider the "causes" that make up the geographic medium, "conditions" that influence the societies that develop in that medium, and clear or diffuse "influences" that are evident and link the medium with the people who inhabit it.

THE RIO DE LA PLATA

In the undefined geography of the New World discovered by Christopher Columbus, the broad basin of La Plata was only slowly revealed to Europe. This Plata would become a high-priority region for those who were stubbornly searching for an interoceanic canal or strait that would allow them to overcome the continental barrier of the Andes, an unexpected obstacle for transportation and commerce on the western route to the Indies and the Spice Islands.

From the moment of its discovery by Amerigo Vespucci (1502), the "river wide as a sea" was ploughed by the anxious seekers of the interoceanic strait and by the ambitious pursuers first of the illusory "mountain of silver"

Translated by Jack Child

Map 8.1   Colonial Political Subdivisions

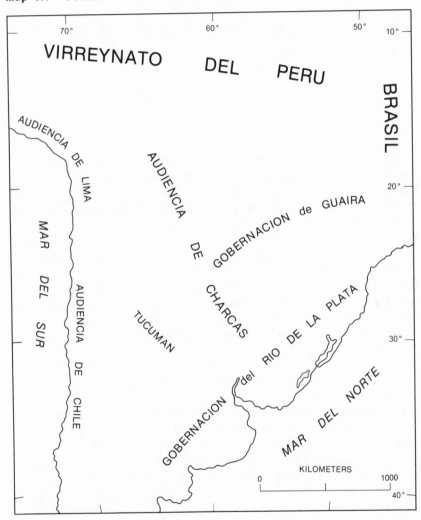

and, later, of the "hinterland" located in Asunción, Paraguay. This maritime space and its coasts, along with the Caribbean (or Antillean) Sea, was a geopolitical and geostrategic area intensely fought over by the Crowns of Spain and Portugal (see Map 8.1).

A year after the discovery (September 25, 1513) of the Pacific Ocean by Vasco Núñez de Balboa, who called it the "Southern Ocean," the Portuguese pilot Juan de Lisboa reached La Plata on a secret trip. This trip was later summarized in the bulletin "News from the Land of Brazil," which was distributed among their clients by the Fuggers, the powerful bankers of Augsburg.

John Schoner, in his 1515 globe of the earth, showed the South American continent cut by a canal or interoceanic passage at the height of La Plata. This led the King of Aragon to send an expedition under Juan Diaz de Solis in 1516, with the mission of drafting the "Official Geographic Map of the New World," later to be called the "Royal Registry of the Indies."

Because of Solis's discoveries, the Portuguese monarch sent two expeditions to the Plata River (or *Mar Dulce*, "Freshwater Sea," as Solis had called it), with the declared purpose of providing for the defense of Brazilian coasts. At the beginning of 1526, Juan III, fearing that a French corsair fleet of ten ships was about to attack his Lusitanian dominions, sent a squadron that founded the first trading posts on the southern coasts of Brazil.

In the La Plata area, other European actors (Holland, France, England) began to make incursions in response to their own imperial geopolitical ambitions. Spain and France disputed the frontier of their spheres of influence in Hispanic-Lusitanian America, which had been marked out by successive treaties, starting with Tordesillas (1494). But the main issue was who would possess both banks of La Plata, because this represented the most important riverine network to be penetrated, conquered, dominated, exploited, and expanded into a vast "hinterland" by means of the great navigable rivers, such as the Paraná and the Paraguay.

With the naming, on May 21, 1534, of Don Pedro de Mendoza as the first *adelantado* (king's representative), the precise limits of Spain's domain were established after a series of notes and diplomatic maneuverings between the king of Portugal, the queen regent of Castille, and Charles I of Spain (who was at the same time Charles V of the Holy Roman–Germanic Empire). This vast region would be called, appropriately enough, the "gigantic province of the Indes." But, in reality, it was La Plata, and its great tributaries (the Uruguay and Paraná Rivers and their tributaries) that geopolitically conditioned the enterprise of populating the region of the La Plata Basin, which today comprises the independent states of Paraguay, Argentina, Bolivia, Uruguay, and southern Brazil.

Toward the end of the eighteenth century, geographic reasoning began to unfold and be disseminated. Up until that time it had been the domain of those closely associated with the state apparatus and the army. They were concerned with practical military and political matters in an epoch in which geographic knowledge was tightly held by the ruling classes and, more specifically, by the princes and military leaders (Y. Lacoste 1959). From Europe emerged a clearer and better defined geopolitical vision of the economic value of the South American market both for the exploitation of important raw materials as well as for the sale of manufactured goods. The significant and growing population concentrated in the La Plata Basin meant a consumer market of real importance for Europe. A brief historical summary demonstrates that for four centuries (from 1500 to 1900), La Plata and its "hinterland" were always present in the expansionist political, geostrategic, and economic plans of the chanceries of the Old World.

The Plata arena still displays geopolitical constants that reveal its importance. The Portuguese-Spanish discovery was motivated by the search for the bioceanic link and the penetration route to the ore riches of the great center of Potosí. The founding of Colonia de Sacramento (in what is today Uruguay) by the Portuguese in 1680, with all the accompanying struggles and treaties dealing with peace and the transfer of territories, was conceived in strategic-military terms for the domination of navigation of La Plata. The Napoleonic invasion of Spain and the fall of King Ferdinand VII permitted the greed of the British Industrial Revolution to acquire dominions in the South Atlantic, enclaves on the coasts of La Plata, and penetration into the commercial field. The historical confrontation between unitarian and federalist Argentines was carried out on both banks of La Plata, with British and French intervention on behalf of the Buenos Aires unitarians, and indirect participation by Italy. The expansionism of the Brazilian empire led to the conquest of the present-day Eastern (Oriental) Republic of Uruguay, with the goal of dominating and controlling the bank of La Plata, which contains its best natural port of Montevideo. By the same token, the Argentine-Brazilian-Uruguayan alliance led in 1865 to one of the cruelest wars in these regions— the War of the Triple Alliance—against Paraguay, which had displayed an autarchic and powerful nationalism ever since it became independent. In sum, La Plata was the permanent scenario for geopolitical disputes from colonial days up to the period of independence, to the point that the peace of the region was constantly threatened.

The Spanish viceregal system was founded on a geographic and economic logic whose efficiency can be demonstrated by the centuries it endured. Its liquidation, stimulated by Britain, converted that geographic logic into geopolitical fatality. In the economic domain, its disappearance disorganized everything: the old interdependence was supplanted by the independence of each of the parties, and this fractured Hispanic America into the aggregate of nations that has lasted to the present (J. J. Arregui 1973).

Furthermore, in the La Plata Basin, there was a certain indifference toward the idea of "nation." Both political as well as intellectual life were characterized by their dependence on international centers of diffusion (Europe and, later, the United States), which impeded any automatic regional action. In the words of Jacques Lambert (1973):

> Due to the fact that the intellectual elite of La Plata was extremely cultured in the midst of illiterate masses, they remained for a long time as a cosmopolitan and alienated aristocracy, more involved in European than local matters.

Throughout this process, one can observe historically and politically how the formal country slowly replaced the real country. This made it clear that disunity would not permit economic interdependence between the component geographic entities. There was no fluid intercommunication, nor integration in depth. The nation rarely had the opportunity to find out what was going on in the state. Commercial activities in the Plata area were organized according to the requirements of the export market. Agriculture was represented by the latifundio, the mining industry was represented by monopolies, and financial capital answered to the dictates of foreign interests.

FROM HEGEMONY TO INTEGRATION

The rivalry for geopolitical space in Latin America, which began after the discoveries by Columbus (and Cabral in Brazil), was inherited by the Brazilian Empire and the new Spanish-speaking republics.

In South America, and in the specific case of La Plata we are considering, Argentina and Brazil maintained a sparse dialogue, with a high degree of rivalry and hegemonic aspirations with regard to the smaller countries. These political and military attitudes validated the thoughts of Raymond Aron (1976): "International actors cease to be acceptable partners if their peoples are separated by painful memories that they do not wish to forget, or by the pain of wounds that remain open." In the case of the La Plata Basin, both Portuguese expansionism as well as the viceregal domination by Buenos Aires changed gradually.

Golbery do Couto e Silva (1967) defined this geopolitical situation, taking into consideration the value of space as limit and launching platform for political action:

> The analysis of space refers mainly to geopolitics applied to the internal domain. This has as its objective total integration and giving increased worth to national territory. At the same time, the assessment of position is of interest, above all, to that portion of geopolitics which is externally oriented, and is characterized by the projection of national space towards the neighboring spaces that surround one.

This concept had been developing in Brazilian intellectual circles, in a slow but persistent manner, since the nineteenth century. In Brazil the concepts of cartography and conquest of territory were born at the same time. Conquest of territory represents the materialization of a geopolitical vision, and cartography involves the balance among the factors of space, human and physical resources, and forms of organization.

Mario Travassos (1947) presents the antagonic thesis with regard to the function of the two great riverine routes of communication in South America: the Amazon and La Plata. With regard to the basin of the latter, Travassos considered that Buenos Aires (because of its higher southern latitude), was in a secondary position with regard to the Brazilian oceanic ports located closer to European markets. These ports, through "export corridors," could satisfy the commercial needs of the landlocked countries of the South American "hinterland" (Bolivia and Paraguay), and could, in this manner, contain the hegemony of the ports of Buenos Aires and Montevideo in the Plata Basin. Behind this geopolitical conception one can find the dictum of Frederich Ratzel (1923): "Space is power."

Internal political conditions in Argentina from 1930 until our times, however, did not permit this thesis of a single spatial concept to evolve. The enlightened Argentine Manuel Ugarte recognized this (1962):

> Far from obeying a logical inclination, Argentina has developed a border policy that in certain cases led it to quarrel with Chile; in other situations it was oriented to prolonging an absurd rivalry with Brazil. It was also constantly ready to debate borders with Bolivia or dispute jurisdictions with Uruguay, and thus wear itself down in secondary actions that, whether it was right or not, reduced its prestige in dealing with the United States and with Europe. By proceeding in this manner, we have forgotten that in diplomatic action, as in military action, there exist strategic fatalisms.

Geopolitical studies in Argentina began to take shape starting in 1940 (J. Atencio, J. T. Briano), although these authors did not study the real and absolute problem of national space in depth. Pragmatic thinking did not emerge for several decades. The foundation stone of Argentine geopolitics was subordinated to the "problem of Brazil," to the effects of the Brazilian–U.S. alliance of the 1960s, to the subimperial thesis of Golbery do Couto e Silva, and to Brazilian projection into the South Atlantic and Antarctica.

It has only been in the past three years, with the new Argentine democracy, that we have seen an increasing awareness with regard to the reorientation of the country in the new South American context. There are symptoms of a return to continentality and to a reencounter with the nation's historical origin:

> History does not forgive failures, and Argentina has failed for a good part of the twentieth century. Surrounded by neighbors who have a much higher demographic growth rate than ours, with vast unpopulated spaces, we must rely on ingenuity and talent if we want to do more than simply develop, and be outstanding. (Villegas 1969)

The Argentine search for identity, the step from not-being to being, is reflected in the pragmatic thinking of the most modern geopoliticians, such as General Guglialmelli:

> The articulation with this America is fundamental for Argentina. But it would not have any purpose if we did not, first, also resolve our own integral development in the total orbit of the nation (1974).

History shows us a series of cycles, including political, cultural, economic, religious, technological, strategic, and geopolitical ones. The La Plata Basin also had its cycles. These include the historical-political ones we have indicated, the cultural and technological ones that occur with greater regularity and intensity from one country to another, and strategic and geopolitical ones.

If we analyze this vast territory of 3,100,000 square kilometers, with a population of over sixty million, and where there is international trade of forty billion dollars, we can see that a new historical cycle has begun. This cycle has been spurred on internally and externally by politico-economic circumstances. A fundamental fact is the return of these countries (except Paraguay) to democracy. This opening up of horizons has led the highest leadership levels to offer to those intellectual, managerial, and labor sectors a frank and pragmatic symbiosis of common interests. This has been accomplished through the most flexible channels of interrelationships, cooperation, and integration of space in the La Plata Basin, and also to all of Latin America.

This new cycle is characterized by a radical change of position: from rivalry to integration, from conflict to cooperation, from individuality to regionalism, from bilaterality to internationalism, from conceptual goals to real achievements, and from competitiveness to societal relationships.

But we speak of the La Plata Basin. There is also a clear vision of the Southern Cone. South American frontiers awaken. The old border demarcations are no longer seen as rigid lines, presumably immutable. The old concept of frontier, which in the language of geopolitics means "the live presence of man," is left to one side, and we carry out true encounters of peoples. If Stendhal could write in 1823, "The world is like a book in which one has only read the first page if one only knows one's own country," then today Latin America is now permanently engaged in reading subsequent pages.

Did the La Plata Basin Treaty of 1969 fail? We believe that the problems identified in a report of an ad hoc group created in the Sixteenth Ordinary Meeting of Foreign Ministers have gradually been solved:

> Stable and permanent concrete forms were not structured for multilateral action, and this meant that the potential for regional cooperation was not realized at the multilateral level. There was a great dispersion of efforts that, despite some attempts at regulation

and organization for cooperation, could not be translated into practical and operational decisions (1976).

No government can avoid the implication that even the most insignificant modifications in important frontier regions in the hinterland of the Southern Cone are "challenges-responses," in the words of Toynbee. These could mean the acceptance of changes in the field of political organization, with effects in economic, cultural, and technological areas as well.

The dynamics of the subregion make it necessary for nations to reformulate their social and economic behavior by reviving their respective national projects, with the goal of designing modern states in the medium and long range. And this implicitly requires the updating of such concepts as nationalism and sovereignty. These must be more elastic, because in this interdependent and multipolar world, both nationalism and sovereignty, with their old nineteenth-century content, have become obsolete. Integration is today another form of shared sovereignty, but this does not mean a loss of the nationality, tradition, values, or identity of each nation.

Another factor that has been considered and supported politically is included in the reality that for each geopolitic there is a geostrategy. In this respect, in less than a decade there has been a true revolution in the physical integration of the Southern Cone.

The agreements signed by the new democratic governments demonstrate that a geographic awareness, which had not existed for centuries, has begun to emerge. This awareness serves as a high priority base from which to understand the whens and the whys of the shared lives and historical development of the nations of the Southern Cone.

Portugal's founding of the Colonia de Sacramento, facing the city-port of Buenos Aires, led it to become the "Achilles' heel" of the La Plata system and its hinterland. This in turn meant that the Plata and the space under its direct influence would become a conflictive frontier on the "rimland" of the Atlantic Southern Cone, to use Spykman's words. It became a shatterbelt between the ruling powers of that age: Spain and Portugal. The reaction did not take long to come. The Spanish crown founded the fort-city of Montevideo, establishing in its port the second naval *apostadero* (staging base) in Spanish America, with responsibility over all the maritime space of the South Atlantic from the Gulf of Guinea to Cape Horn.

Thus, geopolitical and geostrategic responsibility of the La Plata Basin was linked to the importance of the South Atlantic as a communication route with Europe and the United States.

COOPERATION AND CONFLICT IN THE LA PLATA BASIN

In our analyses we strongly advocate studying the geopolitical constants that exist in a given space. In the case concerning us here, the La Plata Basin has been a witness to innumerable geopolitical constants throughout the five centuries since its discovery. Some constants we have already indicated. But in the historical-political process of this century, especially since World War II, the "possibilist" vision of man achieved much: the harnessing of the hydroelectric energy of rivers, the construction of bridges to facilitate the easy communication of peoples, and the improvement of natural ports for the entry and exit of products. The hinterland, and especially landlocked countries like Paraguay and Bolivia in the La Plata Basin, are no longer governed by the deterministic current of geopolitics. South America moved from the theory to the practice of a geopolitical relativism, and has been able to discover norms explaining the ways in which geographic factors act on the policy of states. This knowledge permits South America to orient its development in harmony with geographic conditions. It is no longer a question of modifying an existing order, but rather of creating a new system with long-term stability. To summarize, the relativist theory indicates that there exists an interaction between geography and politics in a relative sense. Geography does not, therefore, absolutely or solely determine political events, nor can politics remain immutable in the face of geographic influences.

The extended controversy carried out within the La Plata Basin treaty took form in various issues that, for over a decade, paralyzed the task of multilateral development, in addition to provoking distrust and diplomatic confrontations.

The Issue of the Paraná River Dams

The domain of the Jesuits, with their theocratic state during the colonial period, was geopolitically called "the ring of the Plata." This was a theater for conflict, first, between the crowns of Spain and Portugal, later, between the empire of Brazil and the Paraguay of the López family, and, finally, between Argentina and Brazil.

In the 1920s, Brazil began the planning of a great system for harnessing the hydroelectric energy of the Upper Paraná River, to be followed later by the Upper Paraguay and the Upper Uruguay rivers. Their objective: to provide Brazil's central and southwestern states with a vast energy potential at the service of important industrial regions, which, in turn, would permit the creation of geopolitical frontier centers. The Upper Paraná is the vital axis of the La Plata Basin. Although the rivers of the basin are born in Brazilian territory, the key to their control is held by Paraguay.

In June of 1966, after the definitive demarcation of the water frontiers along the Paraná River, Brazil and Paraguay signed the Act of the Falls, which approved the construction of the largest hydroelectric dam in the world up to the present. This was Itaipú, with a capacity of 12,600,000 kilowatts, financed by Brazil, Paraguay paying for its share with energy over a fifty-year period.

It is geopolitically significant that Itaipú was constructed barely twenty-three kilometers from the international border with Argentina. Since under international law the Paraná River is considered to have successive stretches of international water, this led to bilateral discussion between Buenos Aires and Brasília, in the first instance, and, later, in the United Nations, and eventually resulted in a regime of prior consultation for the use of its resources.

At the same time, Argentina and Paraguay signed an agreement for the building of another great dam downstream, the Yacyreta Apipe, presently under construction. Argentina is also considering the building of two more dams in its own territory, in the stretch of river known as the Middle Paraná. For the last eight years, Uruguay and Argentina have also operated the binational Salto Grande dam on the Uruguay River, which also belongs to the river system of the La Plata Basin.

When all of these projects are interlinked, Paraguay will be transformed into the Kuwait of hydroelectric energy in the La Plata Basin, and the region as a whole will become the richest in the world in this type of energy, with over forty million kilowatts produced each year (See Map 8.2).

Table 8.1 Hydroelectric Energy in the La Plata Basin

BRAZIL (on the Paraná River and its tributaries within Brazilian territory)	22 dams	8,218,000 kw
Itaipú (50%)		6,300,000 kw
ARGENTINA (on the Paraná River and its tributaries within Argentine territory)	4 dams	6,200,000 kw
Yacyreta Apipe (50%)		2,050,000 kw
Corpus (50%)		2,200,000 kw
Salto Grande (50%)		810,000 kw
PARAGUAY (in its territory)		90,000 kw
Itaipú (50%)		6,300,000 kw
Yacyreta Apipe (50%)		2,050,000 kw
Corpus (50%)		2,200,000 kw
Uruguay (in its territory)		120,000 kw
Salto Grande		810,000 kw

The Mutún Iron Ore and La Plata Geopolitics

The Mutún mountain ridge of eastern Bolivia is barely twenty-five kilometers from the Bolivian-Brazilian border. Its high-grade ore (49 percent, sometimes reaching 60 percent, purity), represents the most important such iron reserves in the world. We should consider that the Soviet Union, in the Korsil region, has reserves calculated as twelve billion tons; in Australia, fifteen billion; and in Algeria–Angola–Liberia–South Africa, fourteen billion

Map 8.2   Hydroelectric Energy in the La Plata Basin

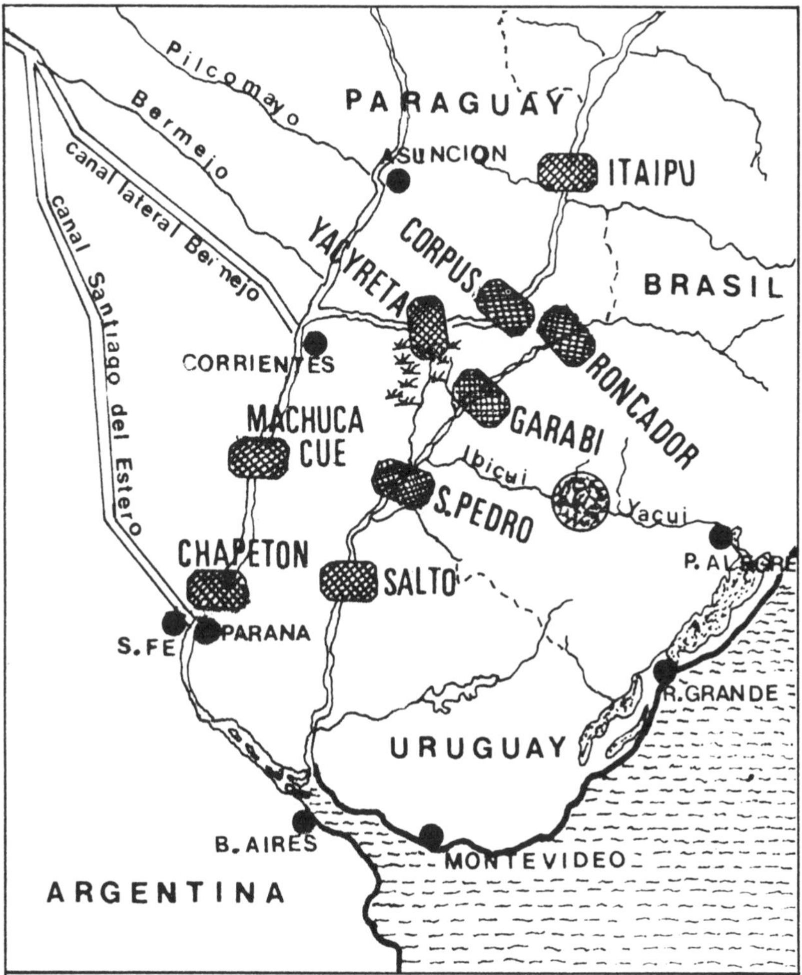

ITAIPU	12,600,000 kw.	CHAPETON	2.300.000 kw.
YACYRETA	4,100.000 "	GARABI	1.820,000 "
M.CUE	3,400,000 "	SALTO GRANDE	1.620,000 "
RONCADOR	2,700,000 "	SAN PEDRO	745.000 "

tons. The Mutún reserves, as calculated by the consultant A. G. Mokee, contain over forty-six billion tons of iron ore close to the surface.

Plans for the extraction of El Mutún ore, and the subsequent creation of a major development pole, have great significance for Argentine interests, as well as for Brazilian ones. This mineral vein, which is related to that of Urucum (in Brazilian territory), will transform the Bolivian area of Santa Cruz de la Sierra into a vital economic and geopolitical crossroads for all of South America, and, more importantly, into a major pole of development for the Southern Cone.

In 1947, the Brazilian geopolitician Mario Travassos (cited previously), defined the triangle formed in Bolivia by the cities of Cochabamba–Sucre– Santa Cruz de la Sierra as an economic space that, under Brazilian influence, would geopolitically neutralize the La Plata Basin. Better said, it would overshadow Argentine iron development, inasmuch as Argentine iron deposits in Sierra Grande, Uchine, and elsewhere are small, inadequate to stimulate a positive development of this industry.

The idea has been taking shape that it might be possible to create an Iron and Energy Community (similar to the Iron and Coal Community of the European Economic Community). Bolivia has the ore; Brazil and Argentina have the technology and the financing; Paraguay has a surplus of hydro-electric energy and timber; and Uruguay has a privileged geotransportational position by virtue of its ports in the basin, which would handle the demands of external markets.

The international accords that the present democratic governments of the region are formalizing demonstrate that in the La Plata Basin there is an expressed political will to create a new shared history. Further, it has become necessary to examine and administer spatial resources, based on a pragmatic geopolitical conception that benefits all the international actors who make up the system.

The Ports of the La Plata Basin

The importance of riverine communications has been well understood by the Brazilian empire, the present Federal Republic of Brazil, as well as the republics of Argentina and Uruguay, and the landlocked nations of Paraguay and Bolivia.

Once it abandoned the thesis of domination of the headwaters of rivers, the Brazilian foreign ministry of Itamaraty gave impulse to a new juridical alchemy: the defense of the freedom of navigation of shared international rivers.

In the 1960s, another geopolitical conflict emerged: the attraction of the Brazilian Atlantic ports for the Plata Basin, to the extent that they would become export corridors for the products of the hinterland. This attraction severely challenged the classical and historical outlet through the great rivers to the ports of Buenos Aires and Montevideo. Brazilian geopolitical analysis sought to neutralize the historical export corridors of the Paraná, Paraguay, and Uruguay rivers toward the La Plata by means of new export corridors along planned roads and river routes, toward the Atlantic ports of Santos, Paranagua, and Porto Alegre–Rio Grande.

All of this complex communications system aimed at overseas ports has been supported with the construction, in less than twenty years, of modern bridges, high-priority highways, links to railroads, subriverine communications, and the establishment of duty-free warehousing and industrial duty-free zones for servicing all of the hinterland.

The geography of transportation has acquired great importance. Some aspects of the possibilism of Vidal de la Blache were used by geopoliticians and planners not simply with development objectives: the application of new conceptual approaches and methodologies has become an important factor for the structural improvement of communications systems in the La Plata Basin. We have seen a double process of assimilation-accommodation and unbalance-balance.

The presidential agreements among Argentina, Brazil, and Uruguay can be considered as the first step toward pushing the La Plata Basin toward becoming a great pole of explosive development in the South American continent. World Bank statistics indicate that this area of 3,000,000 square kilometers will have a population of over 100 million by the twenty-first century. This is a geoeconomic space with a high consumer index, rich in primary foodstuffs and minerals. In addition, the fishing resources of the La Plata and its corresponding jurisdictional waters of the South Atlantic are exploited today under agreements between the basin and the Soviet Union, Poland, Korea, and Spain.

Table 8.2 La Plata Basin Countries

Area: 3,100,000 square kilometers
Population (1986): 80,000,000 inhabitants

Country	% of the Basin	% of the Country in the Basin
Paraguay	12	100
Brazil	45	17
Argentina	31	34
Uruguay	8	80
Bolivia	4	19

LA PLATA BASIN OR SOUTHERN CONE?

The idea of creating a La Plata Basin System (which had its first meeting in the city of Buenos Aires on February 27, 1967), had its origin in the Argentine foreign ministry toward the end of the constitutional government of Arturo Illia, and was carried on by the military government that overthrew him. The system's institutionalization was accomplished in the city of Brasília on April 25, 1969, through the creation of an Intergovernmental Coordinating Committee (CIC, for its Spanish title, Comité Intergubernamental Coordinador), with a permanent seat in Buenos Aires.

The circumstances of these meetings of the foreign ministers of Argentina, Brazil, Bolivia, Paraguay, and Uruguay coincided with the integrationist trends that slowly had been gathering form through bilateral agreements between governments. The trends had also been stimulated by major transgovernmental companies and the industrialized nations. These stressed the thesis of specialization of production by countries, the availability of their natural resources, and the industrial concept of efficiency and economy of scale.

In our spatial concept of southern Latin America, we conceive of the shaping of cohesive nuclei, development centers, and axes of radiation that go beyond the strictly defined orbit of the La Plata Basin. Man-time-space are intimately linked parts and must be carefully considered in making decisions that affect the historical development of a nation.

In the Southern Cone, integration in the economic-social-technological-cultural (and, of course, political) domains has an *internal* as well as an *external* dimension. It has an internal dimension because one cannot break up the interrelationship of the various countries of the area, as was made explicit in Table 8.2. It is essential to avoid demographic imbalances that would provoke a chain of severe economic and social differences between the geopolitical spaces within a given country and, even more so, within the region.

Ratzel (1923) stated that "territory represents in itself a universal and eternal element." And Kirkpatrick (1918) considered that "the Latin American republics should associate with each other according to the characteristics of the space that they occupy."

Insofar as the external dimension of integration is concerned, it should focus on the continental level as soon as the internal and regional integration of each country is accomplished. This statement stems from our belief that, geopolitically and geoeconomically, there exist various Latin Americas, because of the historical, cultural, and social framework, as well as external pressures. These several Latin Americas have not yet been able to find the sustained ties to formalize their true continental nature (See Map 8.3).

FROM DEPENDENCE TO INTERDEPENDENCE

The industrialized powers have consistently attempted to convince the peoples of Latin America, with some success, that the problems of the region were due to its inability to absorb the "modernizing" influences of the

Map 8.3   Continental Zones of Integration

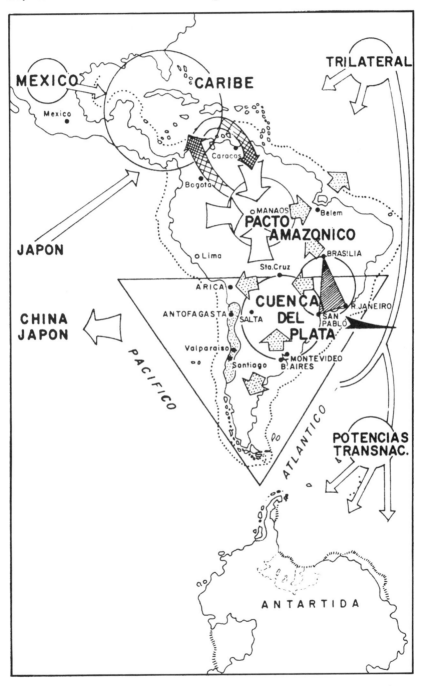

developed nations. They recommended that the region should totally open its doors to foreign capital, to technology without transference, and to institutions and other cultural forms originating in the industrialized world. They recommended that Latin America should have patience, saying that the anteroom of development is underdevelopment, and that this is a stage through which all the technologically developed countries once passed.

In our equation, underdevelopment is not the anteroom of development, but rather the consequence of the development of other powers that were able to act with absolute power. From all this we draw another conclusion: the process of subregional integration within the continental context—such as that of the Southern Cone—should be translated into a liberation project, with the logical, but not utopian, measures that will facilitate moving from dependence to interdependence. As O'Donnel and Link (1973) put it:

> The fundamental strategy of the dependent states is an alliance against the dominant power. To the extent that the dependent nations can overcome the isolation in which the dominant power has placed them, they can then consider a common approach to the resources of power, and with this, introduce a fundamental change in the situation.

It is now very clear that planning is not a mechanical process in the Southern Cone. Planning and geopolitical shaping become mixed in the "alternative futures" a state has. Political leadership must know how to direct—in harmonic terms—the natural processes and the evolution of a society, formulating current projects; for, in the contemporary world, the rational and pragmatic drafting of objectives and purposes is in itself dynamic.

Interdependence in the Southern Cone has consolidated some relationships dealing with exchange of goods and services and division of labor. The integration process in the Southern Cone is subordinated to a global dynamic of *reaffirmation-overcoming*, of *integrative-nationalism* from the subregional to the continental. The country in the region, the region in the continent, Latin America in the world.

A PROJECTION TO THE TWENTY-FIRST CENTURY

Making use of prospective geopolitics—but without getting into the field of futurology—we can conceive of the slow forging of an integrated framework in South America that takes its driving force from the continental Southern Cone.

- *Cohesive nuclei*, in different regions, united by socio-economic axes.
- *Development centers*, some of which begin to radiate geopolitical power toward zones that have important natural and strategic

reserves and that, in certain cases, represent key points to strengthen a service infrastructure at the macroregional level (See Map 8.4).

In this manner, we can foresee the shape of:

1. *Rio de Janeiro–San Pablo–Santos–Belo Horizonte–Brasília–Minas Gerais*, which would formalize a "maneuvering quadrangle," to use the term coined by General Golbery do Couto e Silva (1976). The creation of Brasília broadened Travassos' "valuable triangle," and conquering forces radiated from the center toward the broad, "closed" region of the Brazilian Planalto, uniting it to the development of the Amazon area.
2. *La Plata*, extending its influence to the heartland of its broad basin. We can see, on the one bank, the Argentine megalopolis made up of the Rosario–Buenos Aires–La Plata axis, with links toward the northeast and northwest, where at the present moment the cities of Salta and Corrientes are development centers, with projection toward the frontier. On the other bank of the La Plata, we have Montevideo, with its natural port, its prolongation westward (with a planned deep-water port at the service of the Plata Basin), and its industrial duty-free zones that favor the landlocked countries and areas.
3. *Santiago de Chile–Valparaiso*, which is the only cohesive nucleus in the region beyond the Andes, in a country where both administrative decentralization and regionalization of national space have been applied.
4. *Antofagasta–Iquique–Salta*, which make up an important export corridor in the northern parts of Chile and Argentina, southern Bolivia, the Chaco area of Paraguay, and southern Peru.
5. *Patagonia and Southern Chile*, whose infrastructure permits rapid communication between the ports of the South Atlantic and Pacific, which also have complementary economies.
6. *South Atlantic–South Pacific*, which are maritime spaces with substantial petroleum and fishing resources and reserves of metallic nodules on their submarine platforms.
7. *The Antarctic*, which, although an area of controlled study and development under the treaty (which may be revised in 1991), is a geopolitical region of great significance because of its physical, geological, and meteorological features. From a geostrategic viewpoint in the arena of global power, the Antarctic (together with the southern tip of the continent) stands guard over the Drake Passage, which is the only natural sea-lane of communication between the great oceans of the planet.

To summarize our ideas, we consider that the key problem for the positive projection of the Latin American Southern Cone involves maintaining the pragmatic thrust launched by the present democratic governments. These seek to break false equilibriums and obsolete plans for hegemony, in favor of the construction of a political, economic, cultural, and technological body that will transform all of Latin America into a valid interlocutor in international relations.

Map 8.4   Cohesive Nuclei and Development Centers of Integration

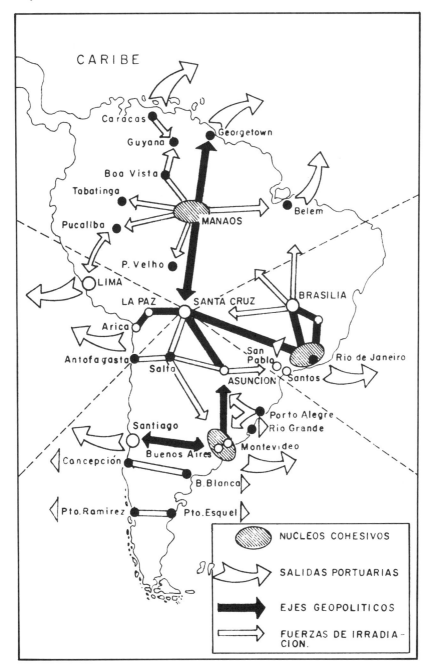

chapter nine

An Energy and Iron Community in the La Plata Basin

Julia Velilla de Arréllaga

The French journalist Bailey stated that the La Plata Basin was called upon to be the "Super Kingdom of the twenty-first century." This is an area of over 3,000,000 square kilometers, with incalculable reserves of iron and manganese and over 120,000,000 kilowatts of hydroelectric energy. In effect, in the Paraná-Paraguay river axis and the Paraná-Paranaiba arc we find one of the greatest potential energy areas of the world.

The La Plata Basin is without a doubt a key area, whose development may be decisive for the integrationist thrust that today is gripping Latin America, inspired by the principles of the men who gave us political freedom. But the advantages are real only to the extent that the countries of the region have the means and the will to exploit these resources.

There have been several attempts in the southern part of our continent to carry out the solidarity-oriented development that in today's Latin America is not merely an option, but an imperative. The creation of The Latin American Free Trade Association (ALALC) in 1960 and of the Latin American Association for Development and Integration (ALADI) in 1980 aroused great hopes. The founding objectives appeared modest, but they turned out to be unreachable, and it has not been possible to create a program to overcome the stagnation. A new prospect has now been created with the agreements signed during 1986 between Brazil and Argentina, with the participation of Uruguay.

Four of the five countries that make up the basin (Argentina, Bolivia, Paraguay, and Uruguay) were once part of the Viceroyalty of the La Plata, which was a geographic grouping created with great geopolitical vision by the Spanish authorities during the colonial period. During the independence period, the five countries of the basin have been actors in the two great international wars that have bloodied the region (1865–1870 and 1932–1935).

Translated by Jack Child

Map 9.1   The La Plata Basin

Despite the hundred years that have gone by since the first of these wars, the people of these nations still suffer the disruptive influence of those tragedies. History shows us what should have been done and what we must avoid. The economic liberation of the countries that make up the basin depends on peace and development. As we overcome past distrust, we need projects that unite us, that identify us, and that awaken hopes for quick fulfillment. The way in which the development of this region is approached will be decisive, not just for the countries located in the great geographic area, but also for all of Latin America.

The Latin American unification process is under way, as can be seen from the new agreements being implemented. Latin America is entering into an era of regional and subregional unions, and the era of continental union is quickly nearing. For the Southern Cone, integration must be achieved first in the La Plata Basin. With cooperation among sovereign powers, no single nation exercising leadership or hegemony (because there will be no true development under tutelage), La Plata Basin can offer the thrust that the Rhine gave to various European countries.

The analysis of this topic led us some six years ago (Velilla de Arréllaga 1982) to propose the formation of an "American Energy and Iron Community" in this region, based on the considerations that follow.

BACKGROUND

Underlying most of the conflicts of recent years is the desire to monopolize the production of iron and energy sources (first, coal and, today, petroleum, hydroelectricity, and many others), as well as the appropriate markets. The old empires, which relied on reserves exceeding their own needs, always attempted to dominate the sources of production in order to control their adversaries by subjecting them to a dependency that made them vulnerable.

The European Ruhr Basin was the cause of permanent conflict. The inequity of the resources of these continental economies, which were frequently rivals, has led to sensitive inequities in development, essentially based on the heavy industries of coal and steel. German power was based on the enormous available energy of the Ruhr, stretched out along the Rhine, and exported in exchange for raw materials through Dutch ports, placing Germany in an advantageous position to rival the greatest industrial states.

The push to create external dependencies began to arouse resistance from neighbors. Clear-minded politicians appealed to reason, which the mindless fanatics rejected, because for them all that mattered was the monopoly of raw materials and markets. Europe had to endure yet another catastrophe that took it to the edge of the abyss before it could pull back and correct its course. At the University of Zurich on September 19, 1945, Winston Churchill, the architect of allied victory, called for Franco-German reconciliation and the creation of a sort of United States of Europe. The Frenchman Jean Monnet accepted that stroke of genius, but suggested that certain preliminary steps should be taken. Thus, on the basis of a memorandum he had drafted on May 9, 1950, Robert Schumann, the French minister of foreign relations, proposed linking German and French coal and steel industries under an independent high authority, and invited the other European countries to join this plan to create a Coal and Steel Community (CSC).

These comments suffice: such concepts could well be transferred to Latin American reality, with a change of degree, but not of form.

Europe started from the beginning: creating the European Coal and Steel Community; this institution smoothed out differences and permitted the formation of a sense of common destiny, security, and shared benefits. The three small countries (Belgium, Holland, and Luxembourg) that played a compensating and balancing role between the more powerful nations, reached notable indexes of progress as their industries developed quickly, generating widespread benefits. On the basis of the obvious benefits of the European Coal and Steel Community, the European Economic Community was created in Rome only six years later, on March 25, 1957. This was a complex but appropriate form of multinational integration, which is a goal pursued today by most of the nations of the world.

Latin America has to recognize that there are no purely national solutions for its problems. The history of a century and a half, and the present reality of regional and subregional organizations, show that only through unity in support of concrete objectives can Latin America have the decisionmaking power to define a common strategy. To achieve this, Latin America must use the means that powerful countries like Venezuela, Colombia, Brazil, and Argentina possess. A community that would take advantage of Bolivian iron and gas, of Paraguayan hydroelectricity, of the riches of Brazil and Argentina, and of Venezuela's incredible potential, would permit us to safeguard our national and regional being, our different ways of life, our right to continue to be ourselves, and our right to be authentically American. If we do not exist as a function of being Americans, we will not exist very long as Venezuelans, Uruguayans, Colombians, Peruvians, Ecuadorians, or Chileans. We will be colonized one after the other, and simply be denaturalized by the empires and subempires that deliberately sharpen our differences in order to convert us into satellites.

We cannot forget what has happened in Europe. These events can be repeated in America, and their memory alone should mobilize our will to avoid them. The conflicts caused by the monopolizing of coal and steel require us to create the American Energy and Iron Community, which would eliminate disruptive factors and create real solidarity.

In an Energy and Iron Community, America would protect its national interests as well as those of workers and entrepreneurs. And rivalries would be absorbed by a foreign policy of prior consultation.

THE SITUATION IN THE LA PLATA BASIN

The difficulties in the La Plata Basin must be overcome. According to geopoliticians such as Vivian Trias (1973, 123): "In this geopolitically conceived space, the key to balance of power is the relationship that either unites, or creates rivalries between Brazil and Argentina."

Besides these rivalries and the friction produced by the development of the region's resources, it is also "evident that concern over their very independence can lead the smaller states to act cautiously before getting too close to a large nation through commercial trade" (Kjéllen 1975, 61).

An attempt to bridge the gaps separating the countries of the La Plata Basin may be politically understandable, but it is historically and technically an error. The desire of the powerful nations of the basin to obtain greater benefits via the monopoly of the region's resources may provoke explosive situations. We repeat that, given these circumstances, our countries must seek the creation in the La Plata Basin of an Energy and Iron Community, which may be the best chance for the small countries to achieve just and legitimate benefits, thus avoiding the repetition of anachronistic imperial procedures.

The iron reserves in the La Plata Basin, especially those of Paraguay (low-grade), of Brazil (in Urucum), and of Bolivia (in Mutún), should not be placed at the exclusive service of the interests of any one nation: an Energy and Iron Community would be an intermediate power. This would avoid the repetition in the La Plata Basin of the kind of procedures the Brazilians endured when they purchased tin in London. This tin was, in fact, ore Bolivia had exported to England, where it was repacked and then reexported to Brazil—as if it were British tin—to supply the Brazilian foundries of the Belgo Mineira company.

Iron is the basic raw material for the "multiplying" steel industry. No country wants to submit itself to the tutelage of another. Therefore, we argue for the formation of a supranational entity to watch over the appropriate development of these riches, for the benefit of the countries concerned and of the La Plata Basin. We are not suggesting an international division of labor under which one set of countries would produce raw materials and another set would produce manufactured goods. What we require is that technology, capital, and manpower be provided in a fair manner.

International experience shows us to what degree the need for raw materials, and especially for the fundamental strategic resources (iron, petroleum, and energy), is at the heart of all conflicts. Monopolistic actions can create pressures and conflicts such as those which, among other things, carried Europe to war. The ideal situation would be for the iron reserves and the production of gas, coal, and hydroelectric energy of Paraguay, Uruguay, Brazil, Bolivia, and Argentina, to be at the service of progress and development of the nations of the La Plata Basin (Velilla de Arréllaga 1975).

Map 9.2   Small Countries of the La Plata Basin

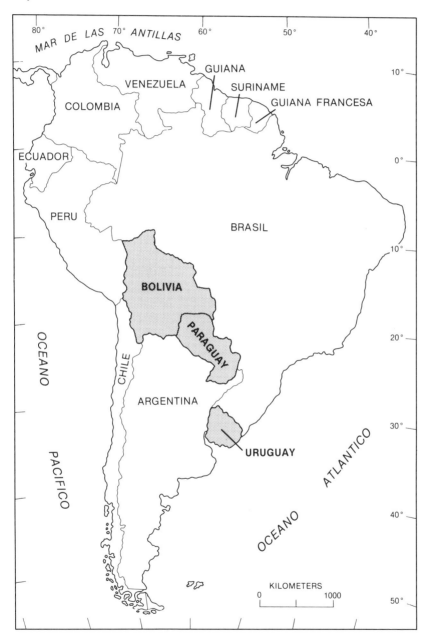

Governments promote different international actions to insure that their national industries have the required strategic resources available, with security and continuity, in peace and in war. This is because the Great Powers know that the role of a few raw materials is of vital importance for national security.

There is abundant data available to demonstrate the importance that our proposed Iron and Energy Community could have. Even in 1975, over a decade ago, Argentina imported 1,282,947 tons of iron: 369,226 of ore and 913,171 of pellets (small spheres of concentrated ore produced using bentonite as an agglutinator). Some 70,000 tons of this ore came from Chile, 11,000 from Bolivia, and the rest from Brazil; all the pellets came from Brazil.

The consumption of steel in Argentina was 4,500,000 tons, which means a per capita index of 175 kilos, a figure higher than the world average. It is estimated that 1982 consumption reached 9,000,000 tons, and in 1985 it was 15,000,000 tons.

Argentina spends at present a million dollars a day on imported steel and related products. What is worse, it depends on a small number of suppliers (Japan provides 60 percent of the semifinished steel products imported by Argentina), so that Argentina is faced with the possibility of suffering a severe industrial strangulation.

Faced with the problem of finding sources of iron, Argentina has used scrap metal in its furnaces, which led a technician to say sarcastically that "if the day comes when we have no more accidents, then Argentina's steel industry is finished." The engineer Humberto San Martín has indicated (1976) that "the output of the Sierra Grande ore deposits in Río Negro Province is no longer sufficient, since they can supply only some 2,000,000 tons of iron ore each year, and in the 1980s we will need at least 9,000,000 tons a year; as a result, some 80% of Argentina's iron will have to be imported."

Brazil's iron production is much higher than that of Argentina, but its civilian and military economy has an Achilles' heel, the supply of petroleum. These situations alone justify and stimulate the need for Bolivia and Paraguay to seek, for the benefit of all the countries, the Energy and Iron Community.

These considerations confirm the need for our countries to step up the process of reaching an agreement to propose the organization of the Energy and Iron Community. Development has a hunger for steel, and this industry, along with the petroleum industry (petrochemicals, and so forth) create the basic and irreplaceable foundation for progress and well-being.

And it is not simply exports outside the La Plata Basin that should provide stimulus; we also have to encourage internal consumption:

> We must recognize that external trade policies are only one part, and
> not the most important part, of a development plan which will

attempt to reach, through high growth rates, a greater national (or regional, we would say) autonomy. This autonomy would include key decisions, and a better distribution of income, especially for the elimination of the worst consequences of poverty (Diaz 1976).

In order to make this export policy efficient, Paraguay must make an extraordinary contribution with its formidable energy potential.

In the Tables 9.1, 9.2 and 9.3, we present data (from various sources) on energy demand and potential that reflect the immense riches these countries possess. To a great degree, this potential has only been partially used, but the present trend is for growing demand that tracks advances in the process of industrial development of the countries involved. This data stems from research carried out for IPEGI (soon to be published) by Efraín Enríquez Gamón, Gustavo Chacón, and Ricardo Canese.

In the light of this research, we believe that the community we have suggested could be part of a valid path toward integration. The sense of common destiny has acquired new vigor, but solidarity so far is passive and horizontal; it has great power, but it is something like the energy potential of a tranquil lake. What is needed is a vertical dimension, in technical, political and economic terms, so that the potential can be converted into energy capable of unleashing forces and possibilities.

The first step could be the American Energy and Iron Community.

Table 9.1 Energy Demand of Countries of the La Plata Basin, 1985

Type of Energy	Argentina	Bolivia	Brazil	Paraguay	Uruguay	Total
Hydrocarbons	37,106	2.042	61,836	508	2,246	103,738
Petroleum	25,250	1,581	59,336	508	2,246	88,909
Natural Gas	11,856	461	2,500	000	000	14,817
Coal	800	000	6,500	000	000	7,300
Others	7,561	546	58,547	363	1,032	68,049
Hydroelectric	5,787	545	51,720	351	1,032	59,436
Nuclear	1,774	000	1,154	000	000	2,928
Alcohol	NA	NA	5,673	12	NA	5,685
Total	45,467	2,588	126,883	871	3,278	179,087

Note: NA=no data available.
Source: Secretaría Ejecutiva para Asuntos Económicos y Sociales, Departamento de Desarrollo Regional, OAS, *Infraestructura y Potencia Energético en la Cuenca del Plata*, Washington: OAS, 1985, pp. 5-9.

Table 9.2 Energy Demand of Countries of the La Plata Basin, as Percentage of Total Demand in Each Country, 1985

Type of Energy	Argentina	Bolivia	Brazil	Paraguay	Uruguay
Hydrocarbons	81.6	78.9	48.7	58.3	68.5
Petroleum	(55.5)	(61.1)	(46.8)	(58.3)	(68.5)
Natural Gas	(26.1)	(17.8)	(2.0)	(0)	(0)
Coal	1.8	0	5.1	0	0
Others	16.6	21.1	46.1	41.7	31.5
Hydroelectric	(12.7)	(21.1)	(40.8)	(40.3)	(31.5)
Nuclear	(3.9)	(0)	(0.9)	(0)	(0)
Alcohol	NA	NA	(4.5)	(1.4)	NA
Total	100	100	100	100	100

Note: NA=no data available.

Source: Table 9.1 above.

Table 9.3 Installed Capacity and Generated Electricity of Countries of the La Plata Basin, 1985

	INSTALLED CAPACITY (MW)			
	Hydroelectric	Thermal	Nuclear	Total
Argentina	4,850[a]	9,078[a]	950[a]	14,878[b]
Bolivia	354	131	0	485
Brazil	35,764	4,196	600	40,560
Paraguay	220[f]	165	0	385
Uruguay	846	593	0	1,439
Total	42,034	14,163	1,550	57,747

	ELECTRICITY GENERATED (GWH)			
	Hydroelectric	Thermal	Nuclear	Total
Argentina	20,557[b]	19,241	5,766[b]	45,564[b]
Bolivia	1,882	354	0	2,236
Brazil	177,980	5,610	3,365	186,955
Paraguay	1,208	180	0	1,388
Uruguay	4,000	357	0	4,357
Total	205,627	25,742	9,131	240,500

Notes to Table 9.3:
a. Installed Capacity, 1983. CIER, *Boletín CIER*, Montevideo, June 1986, p. 2.
b. *Boletín Informativo Technit*, no. 245, January–February 1987, p. 5. For self-generated, *Boletín CIER*, June 1986, p. 2.
c. Secretaría Ejecutiva para Asuntos Económicos y Sociales, Departamento de Desarrollo Regional, OAS, *Infraestructura y Potencia Energético en la Cuenca del Plata*, Washington: OAS, 1985, p. 13.
d. *Mundo Eléctrico*, São Paulo, Editora Técnica Gruenwald Ltda, November 1986, p. 40 and following.
e. ANDE, *Memoria y Balance*, Asunción: ANDE, pp. 9, 11.
f. 30 MW from Itaipú and 190 MW from Acaray. Paraguay ceded 916 MW to Brazil in December 1985. Itaipú Binacional, *Memoria 1985*, Asunción, 1986, p. 10.

Source: From the forthcoming book by Efraín Enríquez Gamon, Gustavo Chacón, and Ricardo Canese, *Energía, Hierro, Acero en la Cuenca del Plata*, Asunción: IPEGEI.

Map 9.3   Energy Sources in Paraguay

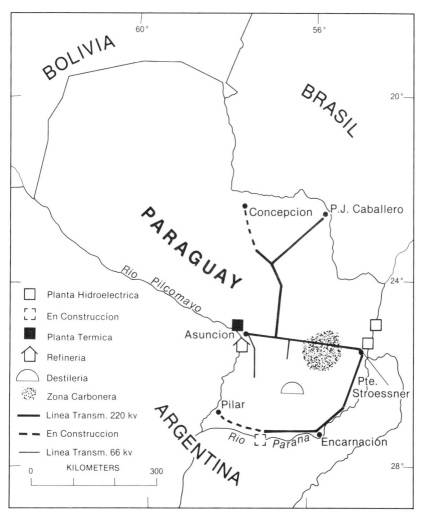

chapter ten

Bolivia's Orientation: Toward the Atlantic or the Pacific?

Martin Ira Glassner

THE COUNTRY

Bolivia is an artificial country, an afterthought, the territory left over after all of the other Spanish possessions on the mainland of the Americas had been liberated. It was not the product of an independence movement; it had no clearly defined boundaries, no distinct character, little internal cohesion, no clear sense of national identity. Even after having lost more than half of its original territory to all of its five neighbors, Bolivia is still a large country, the size of Zambia and Zimbabwe combined, or Texas and California. It sprawls across the high plateau flanked by two branches of the great Andean cordillera, across the intensely dissected eastern slopes (Yungas) and the seemingly endless lowlands of the Oriente, scrub and savanna in the south, tropical rainforest and jungle in the north. Its people are clustered on the Altiplano and in the Yungas, mostly in cities, towns, and small settlements located where wind, water, soil, and slope make settlement both desirable and feasible. The vast Oriente is sparsely inhabited: less than 10 percent of the population lives there.

The grain of the country runs generally north and south, and so do most important transport routes. The only significant exception is the La Paz–Cochabamba–Santa Cruz axis, which could be extended to Charaña on the Chilean border and to Corumbá in Brazil on the Río Paraguay. Bolivian writers, more than others, lament the fragmentation of Bolivia, its continuing lack of cohesion, its regionalism—encouraged, though not dictated, by its topography. This is a country of superlatives: highest,coldest, wettest, driest, longest, emptiest, poorest, least stable, most spectacular, greatest potential, most depressing.

Its centrality in South America is its blessing and its curse. Geopolitical

Map 10.1   Bolivia and Surrounding Areas

Map 10.1   Bolivia and Surrounding Areas

writers frequently characterize it with triangles: Bolivia, spread over the junction of the three great physiographic regions of South America—the Andes, the Plata Basin, and the Amazon Basin; Bolivia, the buffer between the three great powers of South America—Argentina, Brazil, and Chile (and to a lesser extent Peru); Bolivia's hope for eventual solidification— strengthened links among La Paz (in the Andes), Tarija (in the Plata Basin), and Santa Cruz de la Sierra (in the Amazon); Bolivia's heart—Sucre (the original capital), Cochabamba (center of the economically valuable Yungas), and Santa Cruz (the petroleum center, the frontier capital, the focal point of the entire Oriente, perhaps the next capital of the country).

Bolivia's geopolitical orientation is similarly triangular. The western part of the country is emphatically oriented westward, toward Peru and especially Chile. Links with the Yungas are fairly strong, but the lowlands of the Oriente are utterly foreign to most highland dwellers, as remote as Brazil or Europe. The southern part of Bolivia is clearly oriented toward Argentina, with only reluctant ties northward toward the political and economic centers of power. The eastern lowlands have no strong orientation in any direction, though the magnetic attraction of Brazil on the north and east is certainly influential. It has frequently been suggested that the ultimate solution to Bolivia's problems is simply to divide its among its neighbors. While that is unlikely to happen in this era of general territorial stability of states, a separatist movement in the Oriente could well develop, led by the ambitious, hard-driving, independent people of Santa Cruz de la Sierra.

What concerns us here, however, is nothing so grand as this. We simply seek an answer to a question that derives from numerous suggestions, exhortations, and predictions over the last several centuries: Is Bolivia, at last, shifting its national orientation from the west toward the east, turning its back toward the Pacific and facing the Atlantic? We have examples of such shifts in the recent past: Australia from Britain to the United States to Japan and the Pacific islands; the Commonwealth Caribbean from Britain to North America, and, perhaps, now to Latin America; Zambia from the south to the northeast; China from the United States to the Soviet Union to the United States again, to cite only a few of the more prominent cases. Is such a thing happening in the center of South America? We shall see.

HISTORICAL BACKGROUND

High on the bleak and windswept Altiplano, not far from La Paz and near the southern shore of Lake Titicaca, lies Tiahuanaco, home of the seminal civilization of South America, the first major conquest of the people who came up out of the valley of Cuzco to move southward and here acquire the most important elements of their culture. Later, when they had created the great civilization we know as the Inca Empire, they obliterated Tiahuanaco, and it lay undisturbed and virtually forgotten for centuries. Much of it has been reconstructed now, and we can begin to understand something of its primeval glory. Perhaps the local Indian reference to its richly inscribed stone archway, the *Puerta del Sol*, as the *Axis Mundi Americani* is not a fantasy, but a memory. And perhaps not much has really changed from the days of the nameless progenitors of the Quechua culture.

While Pizarro and his colleagues were in Panama planning their assault on the Inca Empire, other Spanish explorers were already working their way up to the Altiplano from the Río de la Plata through what later became Asunción, base for Spanish settlement of most of southern South America. The Spaniards found on the Altiplano what they were looking for—silver, and lots of it. Potosí became the world's largest producer of silver, and a grand city because of it. Not only did settlers come up from Asunción, but for many generations the mules for the mines came from Salta, Argentina, and much of the trade was with the south. But administratively, the Audiencia of Charcas, based in Chuquisaca (now called Sucre) and including Potosí, La Paz, Cochabamba, and so much more, was part of the Viceroyalty of Peru. In fact, it was also commonly known as Alto Peru. It was administered from and responsible to Lima. Ores were exported through Peru. From the very beginning of the colonial period, then, the orientation and loyalty of the people of Charcas were split between Peru and Argentina.

The situation was complicated somewhat by repeated attempts to open trade routes through Brazil via the Amazon rivers and across the Chaco to the Río Paraguay—all were unsuccessful. The problem of divided loyalties was compounded in 1776 when a new Viceroyalty of the Rió de la Plata was carved out of the Viceroyalty of Peru. Many citizens of Alto Peru, of that tiny fraction of the people who knew or cared anything about politics, remained loyal to Peru; others intensified their contacts with Buenos Aires.

The rebellion of the Spanish colonies in America began in Chuquisaca, intellectual and cultural center of Río de la Plata, in May 1809. It was promptly suppressed there and in La Paz, but was successful in Buenos Aires a year later. The new republic was called the United Provinces of Río de la Plata. The loyalists in Alto Peru thereupon withdrew and reincorporated the area into Peru. Guerrilla warfare broke out there and continued for fifteen years, during which time the various guerrilla groups formed six major, and many minor, *republiquetas*. The Argentines invaded Charcas repeatedly from 1810 to 1816, when they finally gave up and turned their attention to Chile. The loyalists retained full control over Charcas until 1820, when the bulk of the Spanish forces were withdrawn to meet the Argentine-Chilean invasion of Peru. The failure of the Argentines to liberate them had turned many people against the United Provinces but not necessarily toward Peru. There was considerable confusion in Alto Peru, with much intrigue, many leaders changing sides and no clear objectives on anyone's part. In February 1824, the Spanish commander seceded from Peru and took over the government, but still no strong movement for independence developed. A year later La Paz was occupied by the forces of the last surviving *republiqueta* and by the combined republican armies that had marched up to the Altiplano after the liberation of Peru. Fighting continued for two more months, and not until July 1825 did a local assembly finally decide, after considerable debate, to create a republic.

The boundaries of the new Republic of Bolivia were quite imprecise, but apparently did include a small portion of the Pacific coast south of the Río Loa. Bolivia failed to obtain Arica from Peru, or even free access to the port, and had to make do with Cobija, which Matthew Fontaine Maury described at midcentury as "an open bay, a miserable town at the extremity of the great desert of Atacama." Not only he, but many others urged development of a transit route through Brazil using some of the Amazon rivers, especially the Madeira. Bolivia under General Santa Cruz pursued an activist policy, thrusting outward in several directions. Navigation on the Amazon and La Plata systems was encouraged and colonization of the Oriente began. Santa Cruz's greatest accomplishment, however, was exploiting the temporary weakness of Peru in 1835 and becoming head of a Peru-Bolivia Confederation. Chile perceived this as a threat, attacked Arica, a "common port" of Bolivia and South Peru, and, after two years of fighting, won the final battle at Yungay in January 1839. The confederation was dissolved and nevermore would Bolivia have any important influence on events in South America.

For the next forty years, Bolivia floundered, trying halfheartedly to develop the Littoral Province and its port of Cobija, while also trying to forge new transit routes through the Beni and the Amazon system, through Tarija to the Río de la Plata and through Chiquitos in the Oriente to the Río Paraguay—in that order of importance. But President Ballivián and his successors failed everywhere. Chile, which had been creeping northward and infiltrating the Bolivian Littoral Province, finally invaded and seized Antofagasta in February 1879, touching off the War of the Pacific (1879–1884). As a result of this war, Bolivia lost to Chile the entire territory between the Pacific and the Altiplano, with its rich deposits of guano, nitrates, silver, and copper, and Peru lost to Chile the port of Arica and the province of Tarapacá.

Thus cut off from the sea, contained behind the Andean ramparts, Bolivia became a landlocked state. Since then, Bolivia has exhibited far more concern for its lost seacoast than it ever did during a half century of possessing it. Bolivia's campaign for some kind of *salida al mar* (outlet to the sea), preferably through Chile, has been the only constant in its foreign policy, waxing and waning with changing circumstances, but never quiet for long. All the while, Bolivia has struggled to find a truly satisfactory transit route to the Atlantic, supporting construction of the Madeira-Mamoré Railroad to bypass over 500 kilometers of rapids and falls; negotiating free port arrangements with Brazil, Paraguay, Argentina, and Uruguay; developing railroads and pipelines through the Oriente; founding Puerto Suárez on a lagoon of the Paraguay in 1877; sending an army eastward toward the Río Paraguay in 1932 and touching off the Chaco War—to cite only a few examples. Yet, for all this activity in the north, east, and south, Bolivia has consistently tilted toward the Pacific. Despite the urging of many and the efforts of a few, Bolivia is still, willy-nilly, a Pacific power. Why should it be otherwise?

WHY SHIFT TOWARD THE ATLANTIC?

Over the centuries, many observers have proposed an Atlantic orientation for Bolivia, for a variety of reasons. They can be summarized briefly under the rather arbitrary headings of politics, geopolitics, economics, and demography. Politically, the argument runs, Bolivia needs the friendship of Brazil and Argentina to counter the hostility of Chile, and these two giants can exert pressure on Chile to grant Bolivia more concessions in transit matters and perhaps even a *salida al mar*. Bolivia also has great geographic advantages: the country is rich in hydrocarbons, hard minerals, and agricultural potential, all of which are attractive to Brazilian and Argentine investors. Bolivia also controls the headwaters of many of the rivers that flow into the Amazon and the Plata, and this fact can be useful to Bolivia in bargaining for various concessions. Bolivia's destiny, say some Bolivian writers, is to join with the Amazon and Plata powers to develop these vast basins, even to benefit from a canal connecting the two drainage systems. There has been considerable exchange of populations between Bolivia and its two giant Atlantic neighbors. Some Bolivians have gone to Brazil to seek work, while northwest Argentina has substantial Bolivian minorities in most of the major cities; meanwhile, Brazilians continue to settle in the Bolivian Oriente.

Most important, with the development of the Oriente, both transit across and trade with eastern neighbors are steadily growing in importance. Argentina and Brazil, in fact, are the major markets for the region's natural gas and some of its oil, and can absorb most of the likely production of iron and manganese from the Mutún mines near the Río Paraguay just south of Puerto Suárez. The rivers and railroads of these two countries provide more direct, hence shorter and cheaper, transportation routes to Europe than the routes through Peru and Chile to the Pacific. And well-developed transit routes to the Atlantic give Bolivia potentially valuable insurance against loss of access to all of the Pacific ports it uses—in the not-entirely-far-fetched event of another war between Chile and Peru.

Bolivia's cultivation of good relations with Argentina and Brazil has paid off to some extent. Argentina not only provides several free zones for Bolivia in Argentine ports, but even a berth in Rosario for the Bolivian merchant training ship, the *Libertador Bolívar*. Argentina supported the nationalization of Gulf Oil, and Bolivia supported Argentina during the Falklands War. Brazil also provides free zones in several ports, of which the most important by far is Santos. Brazil has invested both public and private funds in the Bolivian economy and provided the good offices so instrumental in generating a Chilean offer of a Bolivian corridor to the sea in 1975.

While still cool, relations between Bolivia and Paraguay have improved in recent years. Bolivia is permitted to use the upper Paraguay for transit—when it is possible to do so—and the nearly completed trans-Chaco highway will likely bring some benefits to Bolivia. Relations with Uruguay are good, but, of course, the two countries do not have a common border. Uruguay has provided free zones for Bolivian transit cargoes in Nueva Palmira and Montevideo, but they are little used. All in all, it can be fairly said that Bolivia has generally good relations with its Atlantic neighbors, and that alone is a strong argument for continuing to encourage them and develop even closer association.

WHY NOT AN ATLANTIC ORIENTATION?

While trying to cultivate good relations with Brazil, Bolivia is wary. Brazilian expansionist policies, dating from the first years of Portuguese settlement, have not been jettisoned; they are, in the eyes of many South Americans, simply being carried out more discreetly now. Even some Brazilian geopoliticians candidly refer to Bolivia as one objective of Brazil's continuing *marcha para el oeste*. Other observers see Bolivia as a large potential market for Brazilian goods, as a source of raw materials for Brazil's rapidly growing industries, as a place to invest surplus capital in hopes of extracting substantial profits. In other words, Bolivia is potentially a Brazilian colony, and few in Bolivia can support that notion. Bolivian suspicions of Brazilian motives have engendered great caution among Bolivian policymakers. They have not accepted all Brazilian offers of investment capital, for example; they have held out for better prices for their natural gas, and they have no plans to close the rail gap between Aiquile, east of Cochabamba, and Santa Cruz. Closing this gap would complete the long-dreamed-of Arica-Santos transcontinental railway and would certainly bring some benefits to Bolivia. But it is not currently economically feasible, and Bolivia is suspicious of Brazil's motives in offering to pay for it. Well remembered in Bolivia is the Brazilian penetration into and acquisition of the rubber-rich Acre territory around the beginning of this century.

Similarly with Argentina. In 1932 the Argentine foreign minister,

Saavedra Lamas, boldly declared before the Senate that the government aimed to reconstruct the economic unity of the Viceroyalty of the Río de la Plata. While this statement can be interpreted in several ways, neither Bolivia nor Paraguay can afford to assume that it was perfectly innocent or a mere aberration or an obsolete relic of less enlightened times. Argentina has, in fact, continued to meddle in Bolivian politics. In 1943, for example, Argentina helped General Villarroel overthrow the regime of General Peñaranda, and in 1980 played a major role in the infamous "cocaine coup" that brought the brutal and corrupt General García Meza to power. Furthermore, the long-delayed development of the valley of the Río Bermejo in northern Argentina, that would include canalization of the river for transport across the Chaco, has still not come to pass and may never transpire. Even if it does, Bolivia is unlikely to make much use of it, for the Río Paraguay itself is quite unreliable for navigation of large ships above Resistencia. Bolivia's transit traffic passing through Argentina would probably continue to use the rails between Rosario or Santa Fé and the border.

As for Paraguay, the agony of the Chaco War and the bitterness of defeat and loss of territory are still vivid in Bolivian memory. And Bolivia is still frustrated by its repeated failure to secure a single good port site on navigable water along the Río Paraguay. Puerto Suárez was so unsatisfactory that the port was abandoned years ago. Even a new port nearby would still have to compete with the larger, well-established, better-endowed port of Corumbá in Brazil, directly on the Paraguay. And Puerto Busch, farther down the river and close to Paraguay, is still only a dream. So is Mutún. Although the mines are working on a very small scale now and shipping out ores by truck, it seems unlikely that the plan to supply the Argentine mills at San Nicolás will be realized very soon. Paraguay, unlike its neighbors to the north and south, has little to offer Bolivia, not even a good transit route.

What about the development of the Oriente? Isn't that supposed to shift economic and political power from the Altiplano to the llanos? Wouldn't that almost automatically mean a turn toward the Atlantic? The Oriente is developing, and fairly rapidly, though still not on the scale projected by some nationalistic dreamers not long ago. The government in La Paz has consistently expended great effort and considerable cash to assure that the new settlements in Pando, Beni, and Santa Cruz departments are linked with La Paz, not with Brazil, even if the links are primarily by air and sometimes rather tenuous. Even fresh meat, even live cattle are flown to the Altiplano rather than sold in Brazil.

While technology has changed since 1910, the physical geography of the Bolivian Amazon has not. At that time, Isaiah Bowman, later to become one of the greatest U.S. geographers, reported that the region traded largely with La Paz. Even goods from Europe came through the Straits of Magellan to

Antofagasta or Mollendo, then overland to Oruro or La Paz, thence to the Beni and elsewhere in the north. This route was easier and cheaper than using the Amazon rivers. The Madeira-Mamoré Railroad, built at such a dreadful cost in money and lives, never lived up to its promise and has been abandoned in favor of a motor road bypassing the falls, and no new railroads are even planned in this enormous region that does not need and cannot support them. Bowman listed "a number of convergent factors [that] operate today to hold trade to the old routes to the Pacific," including the difficulties of hauling ores over tropical jungle rivers, several dangerous tropical diseases, sparse local population to provide labor, unwillingness of highland dwellers to come down to the plains, and the costs of building and maintaining the requisite infrastructure. With only the slightest modification, Bowman's assessment is as valid today as it was in 1910.

While a geopolitician could claim in 1965, as did Lewis Tambs, later a U.S. ambassador, that Santa Cruz de la Sierra is "the epicentrum of the continent," that assessment is at best premature, for La Paz is still the "epicentrum" of Bolivia, still the great node of population, government, transportation, culture, and economic activity. Undoubtedly, Santa Cruz faces a bright future. Undoubtedly, its importance will grow considerably, even in the near future. But this does not mean that La Paz will be eclipsed, that the importance of the Altiplano will dwindle, or that, even should it become the capital of Bolivia, Santa Cruz will automatically turn east or south and diminish its relations with the heart of Bolivia to the west.

ATLANTIC LINKS

Generally, as we have noted above, Bolivia's relations with all of its Atlantic neighbors are good, though with some suspicions and problems lurking in the background. Those with Argentina are particularly strong. The two countries have established a Comisión Argentino-Boliviana de Integración Física (Argentine-Bolivian Commission of Physical Integration) to develop closer ties. Bolivia has numerous treaty ties with all four Atlantic countries, covering a wide range of normal activities, and all seem to be carried out with no major problems. In addition, Bolivia belongs (or belonged) to several subregional organizations in which it has played an active role.

In 1963, Uruguay, Paraguay, and Bolivia, under the auspices of the Interamerican Development Bank, agreed to form an organization called URUPABOL, through which they could develop some form of economic integration. It took them fifteen years, however, to hammer out a treaty formally constituting what was an informal organization, and another three years passed before the treaty entered into force. But by that time, Paraguay had lost interest and URUPABOL collapsed in 1983 without having accomplished anything of consequence.

Far more important is the Cuenca del Plata group. Under the terms of its 1969 treaty, all five countries that share the Plata Basin (Brazil, Bolivia, Paraguay, Argentina, and Uruguay) have been working together to develop many kinds of multilateral, bilateral, and unilateral projects in the basin, designed to strengthen all of their economies but without any real integration effort. It is modeled on the Tennessee Valley Authority and coordinated by the Organization of American States, which also provides financial and technical assistance. Many infrastructure projects are already functioning, and more are planned and under construction. Massive hydroelectric dams, new highways and bridges, water supply and irrigation projects, local public works and social services, navigation, and other activities have brought lasting benefits to all five countries. Despite disagreements, delays, inefficiencies, and misunderstandings, it continues slowly but noticeably to tie the countries more closely together.

The Treaty of Amazonian Cooperation was signed in Brasília in 1978 by Bolivia, Brazil, Colombia, Ecuador, Guyana, Peru, Suriname, and Venezuela. Ostensibly, its objective is to encourage the integrated development of the Amazon Basin, but some of the signers are a bit doubtful about Brazil's objectives in initiating the agreement, and not all have yet ratified it. To date, little has been accomplished under the treaty, but there is still interest in it, and it is possible that some day it may bring substantial benefits to its parties. Meanwhile, Bolivia has lost nothing by signing it and has probably earned a modicum of good will in Brasília.

Bolivia belongs to two regionwide organizations that have headquarters in nearby capitals. The Latin American Economic System (SELA) is the successor to the failed Latin American Free Trade Association (ALALC) and inherited its headquarters in Montevideo. The Latin American Railroad Federation (ALAF), sited in Buenos Aires, joins together private and publicly owned railroads that work together to improve rail transport throughout the region. Both, in their respective spheres, are helping to reduce frictions in Latin America and smooth the flow of trade among the members, though neither has been able to stimulate much progress toward economic integration.

Finally, the United Nations Economic Commission for Latin America and the Caribbean (CEPAL) and the Latin American Integration Institute (INTAL) of the Interamerican Development Bank, with headquarters respectively in Santiago de Chile and Buenos Aires, are jointly developing transcontinental transport corridors linking Lima with São Paulo and Buenos Aires. Both of these corridors pass through Bolivia. Since the La Paz–Lima transport routes are already reasonably well developed, most of this development has the effect of bringing La Paz closer to the Atlantic.

One must avoid placing too much importance on these Bolivian ties to the Atlantic, however. Some of them are rather slender and few of them affect many Bolivians directly. Far more important to Bolivia are its Pacific ties.

PACIFIC LINKS

Bolivia, through much of the colonial period and well into the republican period, was oriented more toward the Atlantic than toward the Pacific. The country was repeatedly frustrated, however, in its efforts to obtain transit rights and facilities in Brazil, Paraguay, and Argentina, and faced the physical obstacles already described. For more than a century now, Bolivia has been tied more closely to the Pacific. Some of the links are political, some economic, some psychological. All are important.

Fundamentally, Bolivia is tied to the Pacific coast because it is near its center of population, government, and economic activity, because rail and road routes are better here than in the east, because Bolivia used to reach down to the sea, and because the seacoast towns are very attractive to those people on the Altiplano willing and able to go to them for business or pleasure. Since signing the Treaty of Peace, Friendship, and Commerce with Chile on October 20, 1904, in which Bolivia definitively surrendered its Littoral Province (the Chilean Province of Antofagasta), Bolivia has continually used the *salida al mar* issue as a rallying point for all factions of the Bolivian polity. By now, three generations of Bolivians, even many of the most isolated and humble, have been raised on a steady diet of propaganda about the lost territory, Bolivia's strangulation at the hands of Chile, the justice of Bolivia's claim, the importance of the dream of a return to the sea. It is difficult to get enthusiastic about the distant Atlantic when one's attention is being constantly drawn to the nearby Pacific.

From a more pragmatic point of view, Bolivia is linked to Peru and the other Andean countries to the north by the *Carretera Marginal Bolivariana de la Selva* (Bolivian Marginal Highway of the Jungle—commonly known as *El Marginal*), a road stretching from Venezuela to Bolivia along the eastern flanks of the Andes, a major feat of engineering that may someday actually generate some of the beneficial effects projected for it. Bolivia is also a member of the Andean Group, a subregional grouping within ALALC/SELA designed to accelerate the integration process for its members. It has not been much more successful than other economic integration efforts in Latin America—and for the same reasons—and at present seems to be disintegrating. Nevertheless, Bolivia has obtained a number of important benefits from membership, including assistance in industrialization and improved transportation.

Bolivia may also benefit from a private effort at integration of northern Chile, northern Argentina, Paraguay, southern Brazil, Bolivia, and perhaps even southern Peru. The Grupo Empresarial Interregional del Centro-Oeste

Sudamericano (Inter-regional Entrepreneural Group of West-Central South America—GEICOS) is an organization of business people in these countries dedicated to developing a new east-west transportation corridor focused on Antofagasta. With modest headquarters in Antofagasta, it is well financed, imaginative and well connected in CEPAL and other governmental and intergovernmental bodies. If successful, the project can bring substantial benefits to the people of southern Bolivia at least.

Most important, however, is the existing transport infrastructure. There are still anachronistic, inefficient facilities on the Pacific side: the rack-and-cog section of the Arica–La Paz Railway (FCALP), the steamship segment of the La Paz–Matarani route across Lake Titicaca. On the whole, however, the facilities are normally adequate to handle Bolivian transit trade reasonably well. A modern (1971) Canadian-built roll-on/roll-off steamship carries up to seventeen railway wagons of either Bolivian or Peruvian gauge loaded with tin and other ores from the mines at Matilde on the east shore of Lake Titicaca in Bolivia to the Peruvian lake port of Puno, whence the wagons can be sent down to Matarani in five days. No Atlantic route can compete with that speed at present. Matarani was opened in 1941 and has entirely replaced the nearby former port of Mollendo. It has excellent facilities for both imports and exports. Arica and Antofagasta have recently been modernized and also operate efficiently—generally.

In handling traffic in transit, procedures can often be more important than facilities. Cumbersome customs clearances and forwarding procedures can delay shipments for extended periods, causing all manner of problems. This was often the case for Bolivian goods passing through Chilean ports, especially imports. Since the mid-1970s, however, the situation has been dramatically improved. In 1975 a CEPAL-sponsored import clearing procedure was adopted in Arica. This procedure, called the Integrated Transit System (SIT), allows goods to be unloaded from ships directly onto railway wagons at dockside and dispatched immediately to the Altiplano, the minimal paperwork usually preceding the shipment. The system works so well that it was adopted in Antofagasta in 1978, is being considered for Buenos Aires, and awaits final Peruvian government approval for Matarani. Again, there is no procedure so expeditious functioning in any Atlantic ports.

As a result of these factors, the bulk of Bolivian transit traffic (chiefly minerals going out and general cargoes coming in) passes through Pacific ports. Some recent figures (in 1000 tons) will illustrate the point :

| | Pacific Ports | | Atlantic Ports | | |
	Imports	Exports	Imports	Exports	Source
Average of 1974–1979	323.1	245.5	245.7	122.0	OEA/CEPAl
1981	463.7	144.8	255.6	62.0	ENFE (Bolivia)

Compare these figures with those given by the U.S. geographer W. O. Blanchard for the year 1918:

Percentage of total foreign trade tonnage			
Mollendo	13	Argentina	8
Arica	33	Brazil	3
Antofagasta	36		

(Blanchard 1918, 343–344)

After examining all available information on the subject, Blanchard concluded in 1923, "The emphasis given the plateau-Pacific traffic seems quite likely to remain if indeed, it does not grow relatively greater" (p. 345). Although the gap between the Pacific and Atlantic trade routes has diminished with changing circumstances, it is still very real and very important. Despite the prognostications of assorted geostrategists, publicists, and visionaries, there is simply no credible evidence that this situation will change very soon, certainly not before the twenty-first century has seen a generation pass by. This fact corroborates the remarkable prescience of Isaiah Bowman, who concluded his extraordinary and still fascinating study of Bolivian trade routes in 1910 with this sentence: "Geographic position, the distribution of resources and climate are here equally powerful factors with topography and drainage, and, strange as it may seem, not only make, but will long continue to make the Atlantic slope, not the Pacific slope, the back door to Bolivia" (p. 192).

BOLIVIA THE BALANCER?

We have considered the positions of those who argue that Bolivia will or should or has become an Atlantic power, and of those who insist that Bolivia remains and will continue to be a Pacific power, and have concluded that the time for a forceful shift in Bolivia's orientation from the Pacific to the Atlantic has not yet arrived and is probably at least one or two generations distant. Since we have answered the question posed by our title, we could well terminate our discussion forthwith. But that would leave unheard a large and consequential group of theorists and practitioners of the craft of diplomacy: those who argue for a middle position.

Bolivia, along with Paraguay and Uruguay, has long been categorized as a classic buffer state, a relatively weak country surrounded by powerful ones that, but for the presence of the buffer, would constantly be jostling one another and from time to time try to destroy one another. Bolivia is indeed that, but some writers, generally among the more serious and level-headed of the observers and theorists, carry the notion even further. They assert that Bolivia is—or can be or should be—"a land of contacts," a "balancer" of the dominant powers nearly surrounding it. This implies an activist role for the country, Bolivia reaching out in all directions, manipulating states and situations so as to reduce antagonisms among the larger countries and bring tranquility to the continent. A noble aspiration, perhaps, but entirely illusory.

The tragedy of Bolivia is that, with the one brief exception of the Santa Cruz confederation with Peru, throughout its history Bolivia has been more acted upon than actor. It has always been an object, not a subject, of international developments. Bolivia looms large in the foreign relations of no country but in the strategic planning of several. Geostrategic maps of South America produced in Brazil, Argentina, or Chile almost invariably show arrows penetrating into Bolivia, not emerging from it (Map 10.2 is an example). Bolivia has too many internal problems, is too poor and too weak, is too victimized by the *narcotraficantes*, is too obsessed with the *salida al mar* issue to take an activist role in continental affairs—or to be taken seriously should she try.

There is simply no evidence of any constructive role Bolivia has played in moderating any of the enduring and often bitter rivalries of the continent. Considering only immediate neighbors, Bolivia has nearly always been a bystander in the antagonisms between Peru and Chile, Chile and Argentina, and Argentina and Brazil. Bolivian writers frequently refer to their country as a crossroads, a meeting place, and, indeed, it is in some respects. But in these conflicts between neighbors, Bolivia is simply bypassed; Bolivia is simply irrelevant.

Bolivia's prime objective has been and must continue to be mere survival. Development and progress and justice and peace and self-respect and other nice things, if possible, but at least survival. From time to time the country has enjoyed the services of skillful diplomats who have managed to win various concessions from the country's neighbors and even from distant benefactors. Perhaps the most potent weapon in the armory of a Bolivian diplomat is the unspoken but haunting fear of the consequences of a total Bolivian collapse. The Altiplano, the Yungas, and the Oriente could instantly become what nature and politicians abhor most—a vacuum. Into the vacuum could plunge headlong the Communists, the Brazilians, the transnational oil and mining companies, or other fearsome creatures, with outcomes too horrible to contemplate.

And so Bolivia survives but does not create, reacts but does not initiate, blusters but does not threaten. Notwithstanding the occasional rattle of broken and rusty sabers on the Altiplano, no one seriously expects savage Bolivian armies to come sweeping down to the west to reclaim the Atacama, or to the southeast to reclaim the Chaco, or to the north to reclaim the Acre. Nor, notwithstanding all the theorists' inbound arrows, does anyone seriously

Map 10.2   Geostrategic Arrows Penetrating Bolivia

Geopolitica (Buenos Aires) Vol. 9 No. 27, 1983. p 18

anticipate an invasion of Bolivia by any of its neighbors. Leaving aside the dubious rewards of such a venture, even if successful, we would be wise to inquire along these lines: If Argentina could not conquer Charcas a century and a half ago, could it do so now? No, the likelihood of major armed conflict in the heart of South America is rather remote, about as remote as the full economic integration of the Southern Cone or the rival giants of the region inviting Bolivia to mediate their disputes.

While of necessity it must for some time tilt toward the Pacific, Bolivia cannot simultaneously act as a fulcrum. The country is in such a desperate situation now that survival alone will drain most of its resources. It has little to gain from foreign adventures, even if they are well intentioned. Bolivian geostrategists could best contribute to their country's survival and development if they keep their triangles and arrows within the country's boundaries.

Note: Field work for this study was carried out in Bolivia, Peru, Chile, Argentina, Paraguay, Uruguay, and Brazil in 1983, on a grant from the National Geographic Society of Washington; and in Bolivia, Peru, Chile, and Argentina in 1985, financed partially by a grant from Connecticut State University.

part five

CHILE

chapter eleven

From O'Higgins to Pinochet: Applied Geopolitics in Chile

_____ Howard T. Pittman

Chilean geopolitics is characterized by a number of unique features, although some characteristics, for example, the influences of the historical legacy and national geography, and the adaptation of Western geopolitical theory and analysis to national planning and government policy and action, are shared in some degree with other Southern Cone nations, such as Argentina and Brazil (see Child 1985 and Pittman 1981 for more details). Chile is unique, however, in both its geographical isolation and its peculiar national geography, which has been accurately described as "_una loca geografía_" (Subercaseaux 1940, 1951). Other features that set Chile apart include: the geopolitical vision of the country's leaders after independence; the continuing disputes, dating back to independence, between the advocates of Americanism and nationalism; the unique geopolitical expansion to the north in the past century, coupled with the loss of Patagonia in the same period; the strong concepts of the value of the use and control of the seas and the passages between them; the degree of application of geopolitical principles to government; and, finally, for the past fifteen years, Chile's rule by a former geographer and professor of geopolitics at the Academia de Guerra del Ejército (Army War College), and author of the book _Geopolítica_, Augusto Pinochet Ugarte. The phenomenon of the geopolitician as ruler is truly unique in South America, for, although such geopolitical theorists as Golbery and Backheuser in Brazil and many others in Argentina had strong influence on their governments, none has achieved government leadership.

This essay seeks to examine and briefly explain the uniqueness of geopolitics in Chile and its influence in domestic and foreign policy and action. (For a more detailed review see Pittman 1981, 1023–1494.) Let us begin the analysis with a review of the nation's geography.

UNA LOCA GEOGRAFIA

As a glance at the map confirms, Chile is a long, very narrow country, squeezed between the Andes and the sea except for the island territories and the Chilean Antarctic. In the north, there are excellent ports and the mineral-rich but inhospitable Atacama desert; in the south, the country devolves into a seemingly endless chain of increasingly cold and wet peninsulas and islands separated by innumerable inlets, bays, channels, and canals. The Chilean Antarctic claim is disputed by both Britain and Argentina and is the site of numerous non-Chilean bases. Only in the central valley and coast is there a hospitable environment, and here is where the bulk of the Chilean population is located. This "crazy" geography still presents problems relating to the integration, development, and security of the nation, some of which cannot as yet be overcome by modern technology. For example, the projected Southern Highway, now under construction, will have a gap of about 200 miles caused by icefields which current construction methods still cannot overcome. This geography also means that both ends of the country remain underpopulated, underdeveloped, and subject to external pressures from historical enemies, such as Peru and Bolivia in the north and Argentina in the south. Furthermore, geographical isolation still exists. The land approaches are all through the territory of traditional rivals: the Atlantic approaches are hampered by Argentine intransigence; the Pacific approaches are secure from Argentine threat, but, although Chile does control the Strait of Magellan, the same cannot be said for the Drake Passage between South America and Antarctica. Its continental isolation and peculiar geography have forced Chile to the sea since its inception, and it is perhaps more maritime in nature and orientation than any other Latin America country.

Chile's historical legacy was, in great part, affected by the country's odd geography.

THE HISTORICAL LEGACY

The Spanish Empire recognized the strategic location of Chile and administered it as a separate colony. The captain-generalcy, or kingdom, of Chile was developed as a breadbasket for the mines of Peru and as the guardian of the sea routes through the Strait of Magellan and around Cape Horn with its great fortress at Chiloe, the last part of Chile to be freed from Spain. Incidental to this was the defense of the colony against the warlike Indians, with such high costs to the empire that Chile was called the "Flanders of the Indies." The centuries-old conflict with the Indians also contributed to the training of generations of Chileans in warfare, for, by independence, most of the troops and many officers were native Chileans, an advantage to the newly independent nation in its early struggles.

In the colonial period, the hostile deserts of the north and the resistance of the Indians, transportation problems, and difficult climatic conditions in the south delayed the expansion of the colony, but these factors encouraged the development of a hardy and close-knit nucleus of population in the central valley and coast of Chile, perhaps the most unified colony in South America at the time of independence (Edwards Vives 1953; Eyzaguirre 1978, 15–22).

Spain also recognized the strategic importance of the Strait of Magellan and attempted to establish two colonies there to control the passage in 1584, but they both disappeared without a trace. Following this example, Chile fortified the strait in 1843, thus achieving control of it early in its history as a nation.

Spain also actually constricted Chilean expansion in this period by transferring Tucumán to the Audencia of Charcas in 1584 and Cuyo to the Viceroyalty of the Río de la Plata in 1776, thus limiting Chile east of the Andes to Patagonia south of Cuyo. This Patagonia area was claimed, but never colonized, by Chile, and later became subject to Argentine conquest.

Thus, at independence, Chile was confined to the area between the Andes and the sea, a region called the "Pacific Balcony" by Emilio Meneses, and *Chile Antiguo* by other writers. Chile did, however, possess a strong, independent, and homogenous population trained in warfare, and, after a relatively short period of conflict and anarchy, it rapidly developed into a strong and powerful state and a sea power that made its influence felt in the Pacific. Despite its unity toward outside events, internally, Chile began to alternate between two often opposing concepts, those of nationalism and Americanism, which have affected Chilean internal politics and international relations ever since.

Americanism and Nationalism in Independent Chile

Americanism—a vague idea of hemisphere unity—is usually identified with Bernardo O'Higgins because he marshalled Chilean power and used it to help secure the independence of Peru. Yet O'Higgins also established Chile's first army and navy and the military and naval academies to support them. He also urged the Chilean control of the Strait of Magellan. Further, O'Higgins argued that Chile should extend to Antarctica. These very nationalist ideas and actions illustrate the contradiction of Americanism and nationalism that has influenced the nation's policies throughout its history.

Nationalism—a spirit of separate national identity—is most frequently linked to Diego Portales, who originated the ideas that Chile must always predominate in the Pacific and never allow a union of Bolivia and Peru (Pittman 1981, 1024–64). It is to the ideas of Portales, and the actions of the leaders who followed him and carried out these principles, that the successful Chilean expansion and the victories over Bolivia and Peru in the remainder of the century are attributed.

Actions of subsequent Chilean governments resulted in the occupation of the Strait of Magellan in 1843, over Argentine protests; the breakup of the Bolivia-Peru federation in 1839; and the progressive occupation and seizure of mineral-rich Atacama to the north from 1830 to 1878. This culminated in the War of the Pacific in 1879, which left Chile in possession of greatly expanded and valuable territory at the expense of Bolivia and Peru, leaving the former a landlocked nation. These actions have led to the perhaps exaggerated perception in Argentina, Bolivia, and Peru that Chile is an expansionist power, a view that still influences geopolitical thought and government policies in those nations (Pittman 1981, 1064–92; for a detailed account of Chilean expansion to the north, see Fifer 1972).

During the same period, however, Chile lost claim to Patagonia as a result of the Argentine conquest of the Indians and subsequent occupation of the territory. Revisionist historians (Encina 1949; Eyzaguirre 1963, 1977, 1978; and Espinosa Moraga 1965, 1969) have charged Americanism and Americanist Chilean leaders with the loss of Patagonia, although other analysts (Meneses 1978) argue that Chile simply could not maintain its conquests in the north and fight for Patagonia at the same time. The disputes over Patagonia and its adjacent islands and waters were the subject of Argentine-Chilean agreements and treaties in 1881, 1893, 1892, and 1984. After four treaties and over a century of disputes and controversies, however, these issues may not be finally settled, as hardliners in both countries still object to the 1984 agreement. It is true that, at present, a workable agreement has been achieved. Still, if history is a guide, more controversy could erupt in this area in the future.

In the meantime, development and immigration beginning in the latter half of the nineteenth century had changed the balance of power in the Southern Cone. Argentina in particular had grown enormously both in population and wealth. Chile, with a much slower rate of growth, could no longer muster the relative power it had possessed and used so successfully in the postindependence period. Chile has not been an expansionist power for over a century, yet the memories of the past linger on among neighbors and rivals, and many of the disputes of the past century remain active today, with unceasing Bolivian demands for access to the sea through Chilean territory and continuing Argentine efforts to limit Chilean access to the Atlantic and to dispute its Antarctic claims.

In more recent times, Chile has continued to defend its interests in boundary disputes with rival neighbors, especially Argentina. Chile was the first South American country to make an Antarctic claim (1940), the first to declare a 200-nm economic exclusive zone (1947), and has been a pioneer in the application of geopolitical concepts of land to maritime space (such as the idea of a national Chilean sea, the equivalence of land and sea space, sovereignty over the 200-nm economic exclusive zone, and the idea of the future importance of the Pacific Basin). In the days of the supertanker and the declining capabilities of the Panama Canal, Chile once again can assume an important role as guardian of the southern passages between the Atlantic and Pacific and between South America and Antarctica. With these ideas in mind, we can now turn to the development of Chilean geopolitical theory and the pragmatic use of geopolitical analysis in government policy and action.

CHILEAN GEOPOLITICAL THEORY

It is a paradox that, while Chileans are often perceived and portrayed as consummate geopoliticians by such rivals as Argentina and Bolivia, they have contributed only a limited amount of indigenous geopolitical theory when compared to the highly developed Brazilian school and the contending Argentine concepts for a national geopolitics, or "national project." What is apparent is that early Chilean leaders were instinctive geopoliticians who established, and in many cases achieved, geopolitical national goals in the past century, and that their modern successors have successfully applied geopolitical theory and analysis to government plans, policies, and actions, both internally and externally, to an unprecedented degree. Furthermore, there seems to be a clear progression from the pregeopolitical thought of such early leaders as O'Higgins and Portales to the analyses and commentary of contemporary Chilean geopoliticians and the plans, policies, and actions of the present government.

Indigenous geopolitical comment and theory began to develop in Chile about 1940, with the writings of Ramón Cañas Montalva, who successfully urged the declaration of a Chilean claim to Antarctica and was the founder and leading writer of the *Terra Australis* group.

The Terra Australis *School of Geopolitics*

Most of the early Chilean geopolitical thought and commentary was published in the pages of the *Revista Geográfica de Chile: Terra Australis* during the years 1948 to 1971. Cañas Montalva originated four geopolitical concepts (1948): (1) the coming of the "Era of the Pacific," (2) the importance of geographical location, especially that of Chile, (3) the geostrategic responsibility of Chile for continental defense and its own destiny, and (4) Chile as a South Pacific power. His concepts included the visualization of Chile as a Pacific power within the context of a unified American continent or hemisphere, and he called for the creation of a confederation of the Pacific as the first step in this integration (1949). He repeatedly warned of the danger of the Argentine threat, urged the establishment of a Chilean Pacific strategy, and emphasized the importance of the Beagle Channel to Chile (1953, 1955, 1956–1957, 1960). Cañas Montalva also recommended the internal integration of Chile and the construction of longitudinal transportation routes to the south (1971).

Another member of the group, Pedro Ihl C., also argued (1952) that Chile should propose a commercial union of Pacific powers and work toward hemispheric (Latin American) unity. Ihl publicized the Chilean thesis that the Pacific extended eastward into the Arc of the South Antilles (1951), which would have facilitated Chilean access to the Atlantic. He also popularized the idea (1951) of a Chilean Sea (*Mar Chileno*) extending westward from the mainland to Easter Island, and thence southward to Chile's Antarctic claim, giving the country a vast part of the ocean to control. These ideas and concepts have influenced Chilean policy in the past and, as will be seen later in this chapter, still do so today.

This geopolitical commentary was accompanied by the development of greater interest in geopolitics in the Chilean war academies, particularly that of the army.

Geopolitics of the Academia de Guerra del Ejército

In 1968, professors at the Academia published two books, Pinochet's *Geopolítica*, and Julio Von Chrismar's *Leyes que se Deducen del Estudio de la Expansión de los Estados*. These publications were and are two of the principal Chilean works on geopolitics. Neither is strictly theoretical in nature, rather, both are analyses and critiques of various geopolitical theories and "laws," together with suggestions and recommendations for their use in determining government policy and action, the same concept contained in Argentine commentary. Both works are concerned with the organic theory of the growth and death of states (the state as a living organism), but not in a deterministic manner, emphasizing instead the exercise of human free will in developing and maintaining the state and preventing its death. Although there are frequent references to German theorists, the emphasis is on a strong, unified, well-educated, well-led population, able to overcome the obstacles of geography and make the best of the resources of the state—reflecting the influence of the ideas of the French human geographers as well as Toynbee.

These seminal works are really textbooks that examine a wide range of geopolitical ideas and present them for use by government leaders in conducting the affairs of state. Pinochet, for example, compares the German, British, Soviet, U.S., French, Argentine, Brazilian, and Vatican schools of geopolitical thought in arriving at his own concepts and recommendations, while Von Chrismar examines geopolitical laws from a number of sources. Both books are highly nationalistic, and they urge the creation of Chilean national geopolitics. Since the then Colonel Pinochet is now president of Chile and the then Major, now Colonel, Von Chrismar is the vice-director of the Instituto Geopolítico de Chile (Geopolitical Institute of Chile) and an active teacher of geopolitics in the government academies, their influence remains strong and, in the case of President Pinochet, predominating.

Other Chilean Geopolitical Commentary

In addition to the above, there is one clear example of Chilean Americanist/integrationist geopolitical thought in the writings of Oscar Buzeta (1978), a retired admiral, who expresses a Christian Democratic point of view.

Buzeta traces through history the pregeopolitical and geopolitical thought and actions of Chilean leaders, including those of O'Higgins and Portales discussed above. He identifies and pays tribute to Feire and Bulnes as originators of maritime integration, to Andrés Bello as the father of integration, to President Aguirre Cerdá for his Antarctic claim and as the geopolitician of the integration of national economic space, to Gonzalez Videla (originator of the 200-nm economic exclusive zone) as the geopolitician of maritime space, and to President Eduardo Frei as the geopolitician of continental integration. He urges the development of Chilean national geopolitics emphasizing Christian Democratic principles, regional integration within the continent, and projection toward the nations of the Pacific Basin.

These are the major Chilean works on geopolitics, although Chilean geographies down to the elementary level are written in geopolitical terms, as are the works of the revisionist historians and Pinochet de la Barra's (1976) *La Antártica Chilena*. There are, of course, countless geopolitical analyses, commentaries, and recommendations in such military magazines as the *Revista de Marina, Revista de la Academia de Guerra Naval, Seguridad Nacional*, and in the pages of the *Revista Chilena de Geopolítica*, official organ of the Instituto Geopolítico de Chile (see Pittman 1981, 1256–1328 and 1714–1725 for a review of this commentary through 1981, and current issues of these magazines for further concepts). These articles, however, like the works of Pinochet and Von Chrismar, are policy- and action-oriented; they are not so much theory as analyses and recommendations for the use of geopolitics in government policy and action.

There is one exception: the evolving theory of Emilio Meneses (1978, 1981), who has analyzed the geopolitical projection of Chile and who proposes the geopolitical restructuring of Chile under a new model that would change the character of the nation toward a more maritime orientation to match its geography and location. Meneses conceives of Chile as based on two geopolitical axes, a long north-south axis and a shorter east-west axis, which cross at the Strait of Magellan. He argues that the present geopolitical center of gravity is located just at the base of the Andes. His concept would move this center of gravity westward to the Pacific, increase the strategic east-west depth of Chile in the central region, achieve the balanced and articulated development of the extremities of the country (coastal belt, maritime basins, Antarctica, and Polynesia), and establish the basis for a new foreign policy (see Pittman 1981, 1284–1332 for a detailed discussion of Meneses' ideas). After this brief review of Chilean geopolitical thought and commentary, let us proceed to examine its application to government in Chile.

APPLIED GEOPOLITICS: THE GEOPOLITICIAN AS RULER

In previous sections of this chapter, we have identified and discussed some of the pregeopolitical and geopolitical ideas and actions of early Chilean leaders. Some of these individuals had military backgrounds, others were civilians, but many of their concepts have not only affected Chilean history, but continue to influence Chilean policies and actions at present. In more recent times, civilian presidents from Aguirre Cerdá to Eduardo Frei have been applauded by Oscar Buzeta for their geopolitical actions. Even the controversial Salvador Allende followed the national line in the dispute over the Beagle Channel. What remains is to trace the actions of the Pinochet government since 1973, for these show the degree of application of geopolitics to external and internal government plans, policies, and actions under a geopolitician as president.

First, let us discuss internal geopolitical plans and actions. Probably the most striking internal move by the Pinochet regime has been the complete internal reorganization of the country under a plan known as the "Regionalization of Chile," (CONARA Special Supplement 1979; Canessa Robert 1979). This plan emphasizes the tricontinental nature of Chile, its internal regional integration, and its increasing security and development of Regions I and XII, the most exposed and least developed ends of the nation. Other measures under the plan included revision of regional boundaries, creation of new provinces, and regional/provincial integration and development, for such central planning extends down to the regional and provincial levels. A key idea of this plan is further integration and development to the maximum extent possible within each region in order to improve and strengthen the country.

The regionalization plan was accompanied by a plan to increase the internal integration of the nation, especially the remote southern regions, through the construction of land and maritime highways to the south (portions of which were only reachable over land through Argentina, or by air or occasional coastal shipping). The maritime "highway," consisting of scheduled roll-on/roll-off ferry-service to and from Puerto Montt, began operations in 1980. The land (Pinochet) highway is still under construction, although a number of links have been completed. Ferry service is still required between points uneconomical to bridge. These measures will further integrate the south with the rest of the country, decreasing the isolation of southern population centers and making them more open to settlement and development. The highways will also end the reliance on Argentine roads. (See Pittman 1981, 1357–1364, 1374–1379 and Alvayay Fuentes 1980.)

The regionalization plan was characterized as "pure geopolitics" by President Pinochet (1986) in an interview with the author. The land and maritime highways are just as geopolitical in nature.

Another geopolitical action by the Pinochet government includes the establishment and publication of government goals and programs designed to "make of Chile a great nation." Serving as guides for government action are such publications as the National Objectives of Chile (1974), the National Objective and General Policies of the Government of Chile (1981), and the Socio-Economic Program (1981). They set objectives, and policies to achieve them, under a system of national planning by the Oficina de Planificación Nacional (ODEPLAN), which not only makes plans, but also has a supervisory role in their execution. Typical of this work was the establishment in 1977 of the *Política Oceánica* (Orrego Vicuña 1977), a national maritime policy that confirms Chilean interest in the sea. This was followed by the founding of a Chilean Pacific Ocean Institute (1981).

Education in geopolitics has been furthered by the army war academy's training and certification of civilian professors to teach geopolitics, and the establishment of the Academia Superior de Seguridad Nacional (Academy of National Security) in 1975. This is the Chilean national war college, which has since been renamed as the Academia Nacional de Estudios Políticos y Estratégicos (National Academy of Political and Strategic Studies). Since 1981, geopolitical education has been formalized under the direction of the Instituto Geopolítico de Chile, which has a civilian director (currently Hernán Santis of the Catholic University of Chile) and which includes both civilian and military members. The directors of all of the Chilean government war academies and the diplomatic academy, heads of other appropriate government institutes, and the rectors of major civilian universities are ex-officio members for the terms of their offices. This ensures the participation of these institutions and provides built-in coordination in the education of military and civilian elites. The institute operates under a biennial plan of activities, the *Principios Geopolíticos de Chile* (Reseña 1986). This elite education is complemented by geographic education (down to the elementary school level) that teaches geography in geopolitical terms and by the publication of news and magazine articles that publicize and disseminate geopolitical policies and plans. Official maps produced by the Instituto Geográfico Militar show the reorganization of Chile and include the Chilean Antarctic claim in the national territory.

There are a number of other geopolitical plans and actions of the Pinochet government, but the above should suffice to indicate the degree of geopolitical influence in internal government planning and action.

In external affairs, Chile demonstrates a similar reliance on geopolitical principles. Chilean foreign policy was characterized as "the pragmatic pursuit of geopolitical goals" (Pittman 1984). Nothing in the interim has occurred to change this evaluation. Chile withdrew from the Andean pact when it proved disadvantageous. Chile continues to maintain its Antarctic claim as it has through every government, regardless of its orientation, since 1940. Under the Pinochet regime, efforts to strengthen the Chilean presence and access to Antarctica have continued with such actions as encouraging families to live in Antarctica, and the building of an airfield capable of handling C-130 transports at the Teniente Marsh Chilean base.

In the Beagle Channel case, Chile maintained the same positions and goals under the Frei, Allende, and Pinochet governments. Patient negotiations and brilliant presentation of the Chilean case resulted in the favorable 1977 award that established a boundary in the Channel and gave Chile the islands south of it as far as Cape Horn, thus confirming Chilean occupation dating back to the last century. After bilateral negotiations failed in 1978, Chile quietly mobilized and occupied forward positions against the Argentine threat of war, but without the hysteria that accompanied the Argentine pressure. After papal mediation resulted in the 1984 Treaty of Peace and Friendship, Chile retained land territorial claims, although it ostensibly gave up the rights to a 200-nm economic exclusive zone projected from baselines in the Beagle Channel and the eastern mouth of the Strait of Magellan. This treaty also provides for mediation of future disagreements and greater integration between Argentina and Chile, and it protects Chilean Antarctic claims.

Under the present regime, Chile entered into negotiations with Bolivia (and Peru) over a proposed access corridor to the sea for Bolivia (1975 to 1978, 1987), but neither effort resulted in an agreement. In the first of these negotiations, Chile attempted to arrange an exchange of Bolivian land territory for a land and sea corridor along the border with Peru (Pittman 1981, 1082–1092; 1986), thus creating a maritime boundary out to the 200-nm limit and a demilitarized zone between Chile and Peru, and introducing the concept that land and sea space are equivalent in value and subject to transfer between states. This proposal embodied both geopolitical ideas and goals, as well as precedents involving the equivalence of land and sea space and the transfer of maritime space. The Peruvian counterproposal also would have created a Bolivian Sea, although of greater scope.

Under the Pinochet government, Chile has preserved territorial integrity and maintained its own interests and geopolitical goals, despite outside pressures from neighbors and rivals, such as Argentina, Bolivia, and Peru, and the dislike, if not actual hatred, of the Pinochet regime expressed by Communist, leftist, and liberal forces in the Americas, Europe, and the United States. Chile has not declared a debt moratorium; it continues to encourage free enterprise and foreign investment while increasing both agricultural and manufactured exports, including arms and munition. Trade, long a geopolitical goal, has been increased with other Pacific Basin countries, such as China.

Above all, Chile seeks to maintain an independent foreign policy and as much autarchy as it can to serve its own interests. But when national power is insufficient, the nation joins and supports international associations and agreements that serve to protect those interests. Thus, Chile entered into the Rio Treaty, after ensuring that the hemispheric defense zone included Antarctica, and was a signatory of the Antarctic Treaty, which freezes, but does not eliminate, national claims. Chile was an original member of the Andean Pact, but withdrew when it became apparent that the pact would no longer serve its national interests. Future integration agreements may be expected to follow the same pattern.

We have examined the place of geopolitics in Chile from colonial days through alternating periods of Americanism and Nationalism to the present highly nationalistic, geopolitically oriented regime. The question for the future is whether a return to a democratic government will eliminate or change Chilean geopolitics. The answer probably lies in the type and composition of such a government. The difficulty of prediction is increased by the almost even split among left, center, and right parties in Chilean politics. Thus, any elected government will involve some type of coalition, and its specific composition will be determined at the time. It is almost certain, however, that future governments will continue to pursue geopolitical goals of internal development and of security in internal affairs. The objective of "making of Chile a great nation" will not materially change. In foreign affairs, a return to a Christian Democratic government, by no means assured, would probably result in greater emphasis on the geopolitics of integration, as discussed by Buzeta. But national goals and aspirations will probably continue alternating between Americanism (integrationism) and nationalism in this unique country.

part six

ANTARCTICA AND THE SOUTH ATLANTIC

chapter twelve

South American Geopolitics and Antarctica: Confrontation or Cooperation?

Jack Child

A number of converging circumstances suggest that Antarctica may be a major object of geopolitical attention in South America in the decade to come.

In part, this is because many of the other long-standing foci of geopolitics on the South American mainland have found resolution or have decreased in priority. For example, the historic Argentine-Brazilian geopolitical rivalry, which has been a driving force in South American international relations ever since independence, has now apparently given way to a high degree of cooperation between these two key nations. Strains between Chile and Argentina over the Beagle Channel islands and surrounding waters were presumably resolved in the mid-1980s, thanks to Vatican intervention. And the uneventful passing of the Centennial of the War of the Pacific (1879–1883) has lowered tensions along the Pacific Coast of South America.

Further, the trauma of the Argentine defeat at the hands of the British over the Falklands/Malvinas Islands in 1982 led to the emergence of a strong current of Latin American solidarity. This current has taken in other issues, such as the devastating debt problem and the need for Latin American integration in order to face up to outside pressures. Although probably not as strong or enduring as its supporters insist, there seems little doubt that South American and Latin American cooperation is reaching unprecedented levels.

The Malvinas/Falklands crisis also focused increasing geopolitical attention on the South Atlantic and the chain of southern (austral) islands that seem to link the southern tip of South America to the Antarctic Peninsula in a long curving arc out to the South Sandwich Islands and then back. (See Map 12.1). The Brazilian initiative to make of this vast area a "zone of peace" met with rhetorical support from many sectors of Latin America and the Third World, and also served to reemphasize the strategic and economic potential of this region.

Map 12.1   The Arc of the Southern Antilles

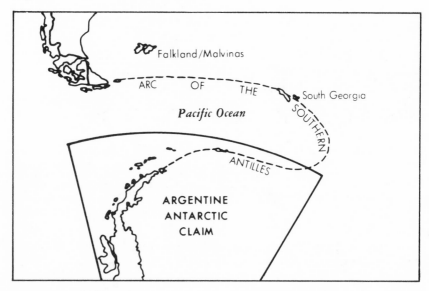

The arc extends from *Isla de los Estados* to South Georgia and the South Sandwich Islands, then back through the South Orkneys to the South Shetland Islands and the Antarctic   Peninsula. Many Chilean geopoliticians argue that the arc forms the natural boundary between the Pacific and Atlantic Oceans. Such a diversion of oceans favors Chile at the expense of Argentina, since both countries accept the bioceanic principle that Argentina is dominant in the South Atlantic and Chile in the South Pacific.

Two other factors account for the increased South American geopolitical interest in Antarctica: resources and time pressures. The resources are more potential and speculative than real and exploitable, but the widely held belief that there is oil and gas on or near the continent is a powerful magnet. To a lesser extent, the proven reality of living maritime resources (fish, krill, seals, whales) is also of interest in a hungry world. The time pressures stem from a clause in the Antarctic Treaty that states that after a period of thirty years (that is, in 1991), any of the consultative members can call for a conference to review the operation of the treaty and possibly call for modifications, including new or expanded claims, or outright withdrawal from the treaty system. This has given rise to a widely circulated misperception that the treaty will end, or must be changed, in the near future. Although this is not what the treaty says, many articles in the popular and geopolitical press take the view that their nation must prepare itself for sudden changes that may occur in the Antarctic Treaty System in the next few years.

Thus, there has been an observable tendency to shift some of the historical geopolitical currents of confrontation and cooperation off the South American mainland to the southern islands and the Antarctic continent beyond. On the Antarctic continent itself, a bewildering array of nations and interested parties have increased their presence in the past few years. There are states that claim large chunks of Antarctica as if they were sovereign territory, while others argue that they have grounds for similar claims, but refrain from making them for the present. There are now almost forty members of the Antarctic Treaty System, but these represent a range of interests. Third World nations, both from within and outside the treaty system, argue that any economic benefit derived from Antarctica should be "for the common benefit of mankind." Meanwhile, various ecological groups advocate keeping the continent a "world park," forever preserved for science and a limited number of tourists.

The Antarctic Treaty System, that unique international political regime that stemmed from the scientific cooperation of three decades ago, now is no longer a "closed" club of the small number of nations that explored and studied it in the distant past.

The original twelve signatory parties (known also as "original consultative" parties because of their decisionmaking role in the system) were Argentina*, Australia, Belgium, Chile*, France*, Japan, New Zealand, Norway*, South Africa, the Soviet Union, the United Kingdom*, and the United States. (The asterisks indicate the seven countries that had staked territorial sovereignty claims prior to signing the 1959 Antarctic Treaty.) Under the treaty, no signatory nation's claim is either denied nor affirmed; no new claims can be made.

Later consultative parties were Poland (1977), the Federal Republic of Germany (1981), Brazil (1983), India (1983), the People's Republic of China (1985), Uruguay (1985), Italy (1987), and the German Democratic Republic (1987). There are thus now twenty consultative (or full decisionmaking) members of the Antarctic Treaty System.

There are also now eighteen acceding parties, which have a voice, but no vote, in treaty system deliberations: Czechoslovakia (1962), Denmark (1965), Netherlands (1976), Romania (1971), Bulgaria (1978), Peru (1981), Papua New Guinea (1981), Spain (1982), Hungary (1984), Sweden (1984), Finland (1984), Cuba (1984), South Korea (1986), Greece (1987), North Korea (1987), Austria (1987), Ecuador (1987), and Canada (1988). These add up to a total of thirty-eight treaty signatories.

It is interesting to note that the three principal geopolitical actors in South America (Argentina, Brazil, and Chile) are all full-fledged members of the system, while other Latin nations with an interest in Antarctica (Uruguay, Peru, Ecuador, and Cuba) are also members. Of these nations, Argentina, Chile, Brazil, and Uruguay all have permanent year-round bases in Antarctica, while Peru and Cuba have consistently sent observers to work and live at the bases of nations friendly to them. Peru now appears to be seriously considering mounting a permanent base, and there is increasing interest in Ecuador in such a project (Mercado Jarrín 1984). Both countries have sent short-term expeditions in summer seasons.

The growing South American geopolitical attention paid to Antarctica and its surrounding waters could acquire either a confrontational or cooperative tone. The history of geopolitical tensions in South America, on the one hand, suggests the likelihood of confrontation, especially among Argentina, Chile, and the United Kingdom. To these historical geopolitical rivalries in Antarctica one must add the newcomer Brazil, and thus the possible transference of the old Argentine-Brazilian geopolitical rivalry to Antarctica (de Castro 1956; Menezes 1982; Pinto Coelho 1983).

On the other hand, the gathering strength of cooperative geopolitics in South America in the mid-1980s suggests that the South American nations may instead take a coordinated and integrated approach. This current has always existed in South American geopolitics but has tended to be overshadowed by the stronger conflictive current. If it is linked to Latin American integration ideals, trends toward redemocratization, and resentment against perceived "colonialist" and exploitative outsiders (the United Kingdom and the United States), this cooperative South American approach may not bring peace and tranquility to the area. It would have to face the firm opposition of the numerous other countries with a presence and interest in the American Antarctic Quadrant running from the Greenwich Meridian to 90 degrees west. These countries now include the United States, the Soviet Union, Poland, the People's Republic of China, South Africa, India, and many others.

THE GEOPOLITICAL SIGNIFICANCE OF ANTARCTICA

Perceptions of the geopolitical significance of Antarctica are a function of distance. There is little geopolitical and strategic analysis of Antarctica in the United States, for example, where the interest focuses more on the scientific and ecological aspects. Those nations closest to Antarctica, especially Chile and Argentina (but also, to some extent, Australia and New Zealand) do emphasize these geopolitical aspects, sometimes to the point of obvious exaggeration. There is also a tendency to make certain parallels with strategic aspects of the Arctic, despite some evident differences.

A region that receives considerable attention is the Drake Passage between the tip of the Antarctic Peninsula and the southern portion of South America. The South American geopolitical literature stresses the great strategic significance of this 600-mile choke point, especially as an alternative route to the Panama Canal, if this should ever close or be denied to a particular nation. This particular type of analysis tends to ignore that there are a number of other means of transportation, and that controlling the Drake Passage is no easy matter because of the distances and difficult weather involved. The significance of the Antarctic Peninsula for power projection into South Atlantic sea lanes is also frequently stressed, but this, too, ignores the reality that the major South Atlantic sea lanes, primarily the oil routes that bend around the tip of South Africa, are very far from the Drake Passage and the portions of Antarctica near South America.

South American geopolitical analysis in this area has tended to stress the significance of the islands between South America and Antarctica. These can be grouped into three categories: the southern South American islands, the islands of the Scotia Arc, and the Malvinas/Falklands. The southern South American islands have always affected Argentine-Chilean relations, but the period of greatest tension now appears to have been surpassed, with the resolution of the Beagle Channel islands dispute. However, this settlement was not well regarded by many nationalistic Argentine geopolitical thinkers, who feel that the solution favored Chile too much, and that Chilean possession of the islands interrupts the continuity between continental, insular, and Antarctic Argentina (Levingston 1985, 28–29). The islands of the Scotia Arc (South Georgia, South Sandwich, South Orkney, and South Shetlands) have geopolitical significance in terms of both possession and their value as markers of the boundary between Atlantic and Pacific. The country that possesses them is clearly in a better position to strengthen its Antarctic claim and to deny the claims of others. (At present Great Britain holds South Georgia and South Sandwich; because the South Orkney and South Shetland islands are within the Antarctic Treaty area, effective sovereignty over them is ambiguous.) Furthermore, if the nation's logistical bases are distant, as in the case of Great Britain, the Scotia Arc islands can play an important role as intermediate support and staging areas for Antarctic activities. The Falklands/Malvinas have had an especially high geopolitical profile since the 1982 Anglo-Argentine conflict, and have long been an important element in Britain's Antarctic projection. Argentina does not need them to facilitate Antarctic activities, but if the country could deny them to Great Britain, it would considerably strengthen Argentina's Antarctic claim. A recurring theme in Argentine geopolitical literature (and, to some extent, in that of other South American countries) is that the North Atlantic Treaty Organization (NATO) is keenly aware of the strategic value of the Malvinas/Falklands in terms of Antarctica, the Drake Passage, and South Atlantic sea lanes. The argument presents the thesis that U.S. and European support for Great Britain in the war was due primarily to NATO's interest in having a base on the islands (Reimann 1983, 56). The presence of many NATO nations in the American Antarctic Quadrant serves to confirm the suspicions of those inclined to support this concept. There are many references to "Fortress Falklands" in the South American geopolitical literature, with parallels being drawn to Gibraltar. Although, in fact, Great Britain has not been able to convince any NATO allies to participate in the construction or manning of "Fortress Falklands," and the forces involved are a drain to NATO commitments, South American geopoliticians remain distrustful.

The renewed attention to the South Atlantic and Antarctica after the 1982 conflict has revived interest in the Antarctic implications of the South Atlantic Treaty Organization (SATO), the Inter-American Treaty of Reciprocal Assistance (Rio Pact), and the South Atlantic Zone of Peace. SATO was initially floated in the 1960s, primarily by South American naval geopoliticians, as a way of filling the strategic vacuum they perceived to exist in the South Atlantic. Although the United States showed some interest, the idea of a SATO never came to fruition because of the distances involved and because, to function effectively, it would have had to include South Africa. The Rio Treaty is relevant because its boundaries extend east as far as the South Sandwich Islands, and south to the Pole. Argentina has always seen this as its geopolitical sphere of influence and has reacted with concern to Brazilian geopoliticians who speak of their nation as having a special security responsibility in the area as far south as Antarctica. In fact, much of the Brazilian geopolitical justification for a growing Antarctic program relies on this type of analysis. It also relies on the sometimes bluntly stated argument that Brazil must protect this area since Argentina has clearly indicated it cannot (de Castro 1983, 29–34). Running counter to this irritant in Argentine-Brazilian relations is the Brazilian initiative in the United Nations to declare the area a Zone of Peace, and to eventually prohibit nuclear weapons in the region. Argentine geopoliticians reacted somewhat ambiguously to this proposal. Although it favors Argentina by placing restrictions on British and NATO military activities, it also represents another Brazilian intrusion into an area Argentines have traditionally regarded as within their sphere of geopolitical influence.

The Arctic parallel shows up occasionally in the Southern Cone geopolitical literature in terms of the transfer of strategic concepts from the Northern hemisphere to the Southern. Some rather imaginative articles have appeared that argue that the superpowers have designs for installing advanced weapons in Antarctica, and accompanying maps show the ranges of various types of nuclear-tipped missiles based in Antarctica (Inter-American Defense College 1985, 10–65). From a realistic military perspective, there is little purpose in placing such weapons in Antarctica because of the physical and political transparency of the environment. Perhaps even more significant, other weapons systems, most notably the nuclear submarine, make the installation of such weapons in Antarctica obsolete. There are some superpower strategic interests in Antarctica, but these are relatively minor, and have to do with communications and data gathering for scientific and missile-launching purposes.

South American geopoliticians have also stressed the significance of transpolar air routes, which is also a concept carried over from the North. It is true that air routes over Arctic regions have considerable importance and have shortened a number of transportation links. As a result, they have drawn increased attention to the region and served to develop it to some extent. Unfortunately, the same cannot be said for the southern air routes in the foreseeable future, because the demand for travel between the southernmost nations of the Southern Hemisphere is minimal. Thus, plans for establishing air support facilities in Antarctica to facilitate flights from South America to Australia and New Zealand have floundered.

A number of geopolitical analyses of the meteorological and ecological implications of Antarctica also seem to be unrealistic. It may well be true that Antarctica influences the climate of the Amazon and the creation of such important phenomenon as the current of El Niño in the Pacific Ocean, but it is not clear what geopolitical significance this may have.

Issues of resource geopolitics have a more realistic base, but, even here, the prospects for any economically exploitable resource must be tempered by the awareness that the resources have not yet been proven. Even if they are, their commercial exploitation must await technology yet to be invented and world prices that would have to be much higher than the present ones.

None of these arguments based on military or economic reality has persuaded the South American geopolitical writers to diminish their interest in Antarctica and the surrounding waters and islands. The presence of the superpowers and their allies in the region, and especially on the nearby Antarctic peninsula, stimulates their interest and suggests to them that something important must be there, or these major powers would not be troubling themselves. There are suspicions that the Soviet Union and the United States are busily prospecting for oil and precious minerals under the guise of science. Even presumably purely scientific projects, such as investigating the ozone "hole" or collecting data to support space exploration, are suspect. In the final analysis, the key variable of geographic proximity continues to exert its geopolitical magic: the nations that are close to Antarctica (and this is especially true for Argentina and Chile) are unwilling to relinquish the feeling that they have a special role in "their" Antarctic territory, and that the presence of outside nations, no matter how strongly supported by history and science, is going to come under close scrutiny. South American geopolitics also has a strong tradition of futurology, looking ahead to the day when technology and the petering out of resources elsewhere on the globe may fully justify all their interest in the resources of Antarctica and the southern oceans.

NATIONAL GEOPOLITICAL APPROACHES TO ANTARCTICA

Argentina

An understanding of Argentine geopolitical approaches to Antarctica is fundamental because of that country's long Antarctic history, and because Argentine Antarctic geopolitics tends to shape the Antarctic policies of several other South American nations. There is a strongly developed "Antarctic consciousness" in Argentina, and a deeply held belief that the nation will never be complete until the various parts of Argentina (South American, Insular, Antarctic, and the Argentine Sea) are under full Argentine control.

Argentine geopolitical analysts stress the strategic implications of the proximity of Antarctica, and Argentina's special role (along with Chile) as guardian of the Drake Passage from Atlantic to Pacific. This interest has naturally tended to focus on the Antarctic Peninsula (called *Península de San Martín* by the Argentines), site of the principal Argentine bases (Fraga 1979).

In the drive to strengthen its Antarctic position, Argentina has taken innumerable steps to increase its presence and undertake administrative acts that might someday be useful in defending its claim. Over the years, these activities have included scientific observations; operation of radio and postal stations; the establishment of colonies with families; the births of Argentine citizens in Antarctic territory; the maintaining of a civil registry for births, deaths, marriages, and other events; and the inclusion of geopolitical arguments for sovereignty in the national school curricula and popular media.

The Malvinas/Falklands conflict served to renew Argentine geopolitical interest in the south, and yet, oddly enough, has seemed to diminish some of the intensity of feeling regarding "Argentine Antarctica." There seems to be a growing realization that making good an Antarctic sovereignty claim is not very realistic and may alienate a number of important allies whose support is needed on the Malvinas issue. Thus, there is cautious but intriguing discussion regarding the sharing of Antarctic sovereignty with other Latin American nations under Argentine leadership (Escudé 1984, 86–89, 160–165; Rozitchner 1985, 121–122; Leal 1983, 14–17).

This cautious, tentative flexibility on Antarctic geopolitical issues is also a reaction to a series of challenges to Argentine interests in the Antarctic. Chile poses one set of challenges, with its own equally long Antarctic history and presence and powerful sovereignty arguments based on propinquity and *uti possidetis*. Chilean-Argentine geopolitical rivalry has had an important impact on their sometimes strained international relations, especially on border issues, such as the demarcation of the frontier along the Andes, possession of the Beagle Channel islands, and drawing the dividing line between Pacific and Atlantic Oceans. It should, therefore, not be surprising that this rivalry shows up in the Antarctic as well. And yet there are also currents of geopolitical cooperation with Chile, most notably the spirit of a 1948 bilateral agreement in which the two countries mutually recognized their Antarctic interests and sovereignty rights, although they were not able to reach any agreement on the limits of their respective overlapping claims. The settlement of the Beagle Channel islands issue in 1985 opened the door for possible greater cooperation in Antarctica, especially if it were to be framed in terms of Latin American solidarity as a means of blocking outside penetration of South American Antarctica.

A different set of geopolitical challenges is posed by Great Britain. The effective British control of the Falklands/Malvinas, South Georgias and South Sandwich Islands has clear implications for British Antarctic interests. As a result, one basic Argentine objective in attempting to obtain these islands from Great Britain is to strengthen Argentina's Antarctic claim at the expense of Great Britain's. On two other archipelagoes within the Antarctic Treaty limits (South Orkneys and South Shetlands), Argentine and British installations exist almost side by side, along with those of several other nations. This is a tribute to the effectiveness of the Antarctic Treaty System, but it is also a warning of some of the implications of a possible breakdown of the system. Argentine concern over British Antarctic and South Atlantic challenges also includes the disquieting possibility of an informal alliance or understanding between Great Britain and Chile, who could be said to share a common geopolitical adversary in the shape of Argentina (*New Statesman* 1985, 8–10).

Brazil presents a more recent geopolitical threat to Argentine Antarctic interests. The Argentine-Brazilian geopolitical rivalry is, of course, a historic one, going back five centuries to strains between Spain and Portugal. But it has only been in the past few years that Brazil has taken an active interest in Antarctica and, in so doing, has undercut Argentine possibilities in that region. The intriguing facet of Brazil's challenge is that it grew out of a purely geopolitical concept: the idea of dividing up the South American Antarctic Quadrant into "frontage" sectors. As shown in Map 12.2, these would be derived from the open (that is, unobstructed) meridians that six South American nations could project to the South Pole. The net effect of this frontage approach is to severely cut into the Argentine and Chilean sectors by awarding a large sector to Brazil and lesser sectors to Uruguay, Ecuador, and Peru. The idea is especially appealing to Brazilian geopoliticians because it undermines the Argentine sovereignty claim (as well as the Chilean), and suggests that Uruguay, Peru, and Ecuador might also have Antarctic possibilities if they support Brazil and not Argentina and Chile. It is doubtful that Brazil is seriously suggesting that these six nations might have sovereignty claims, but rather is simply trying to weaken the Argentine-Chilean sovereignty position by bringing in other Latin American nations. Thus, the end result of this Brazilian geopolitical approach may be to strengthen the Latin American condominium idea under which Brazil

196   JACK CHILD

Map 12.2   The Brazilian "Frontage" Theory

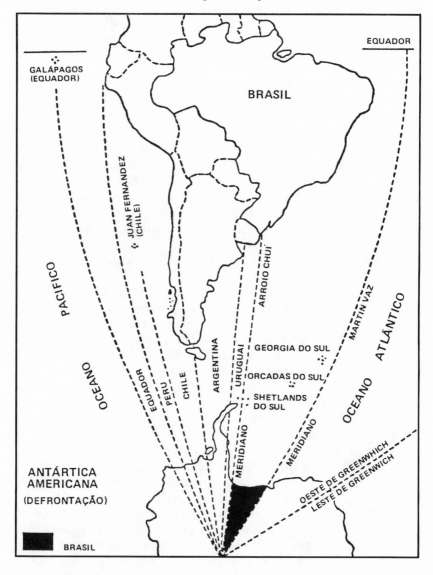

would play a major role as the emerging regional power. The numerous recent bilateral agreements between Argentina and Brazil suggest that a new, unprecedented era of cooperation between these two countries may be beginning. If so, Antarctica may be one arena for this cooperation at the expense of Argentina's territorial claim.

Chile

Geopolitical thinking in Chile has had great influence in many domains of government and the private sector, especially under the military regime installed in 1973. President/General Augusto Pinochet is Chile's pre-eminent geopolitician, and under his administration Chile's Antarctic policy has been strongly conditioned by geopolitical principles. This recent emphasis is not, however, the starting point for Chilean Antarctic geopolitics. Historians would argue that, since the days of independence, Chileans have been forced by their unique geography to be very sensitive to the limitations and possibilities of their space, and that this has naturally led to an important role for geopolitics in national affairs.

Like the Argentines, the Chileans base their Antarctic interest and sovereignty claim on a number of arguments, including several with a strong geopolitical component. Thus, there is a concept of a tricontinental Chile (South American, insular, and Antarctic), tied together by a "Chilean Sea" in the South Pacific. Overlapping Antarctic claims with Argentina, the Beagle Channel issue, and the problem of whether the Atlantic-Pacific dividing line follows the Scotia Arc have all led to geopolitical tensions with Argentina (Marini 1984; Leoni Houssay 1984, 5–20).

Although both countries are obviously the closest Latin American nations to Antarctica, Chile beats out Argentina because of possession of Cape Horn and Diego Ramírez Islands in the Drake Passage. Thus, a consistent theme in Chilean geopolitics is its special status as "the southernmost nation in the world." Chile's first geopolitical journal was appropriately called *Terra Australis* (Southern Land) and contained numerous articles stressing the need for the nation to be more concerned with its Antarctic interests.

The matter of geological and oceanographic continuity between South America and Antarctica via the Scotia Arc is another major theme in Chilean geopolitical writings, and it is presented in such a way as to support the concept of the Scotia Arc as the natural divider. This effectively undermines the Argentine position and strengthens the Chilean. Significantly, the Chilean Antarctic claim, unlike the Argentine, does not have a northern limit. The reason, according to Chilean geopolitical writers, is that there is a continuity between the three Chiles that does not require or permit such a boundary (Marull Bermúdez 1978, 27–34; Cañas Montalva 1979, 89–118).

Chilean geopolitical thinking has a strong maritime component. Independence was partly won by naval battles, as was the major postindependence achievement of the nineteenth century, the successful War of the Pacific against Bolivia and Peru. Toward the end of the century, Chile was the major naval power in the subregion, to the point that its leaders referred to the southeastern Pacific as "a Chilean Lake." This maritime interest has also focused on the three interoceanic passages between mainland South America and Antarctica: the Strait of Magellan, the Beagle Channel, and the Drake Passage. Chilean geopolitical doctrine argues that Chile is the natural "guardian of the doorway" between Atlantic and Pacific and that,therefore, these three interoceanic passages must, to the extent possible, be in Chilean hands. Holding the other side of the doorway (that is, the tip of the Antarctic Peninsula) strengthens the validity of this argument.

Like Argentina, Chile has a well-developed program of Antarctic activities that include science, administrative acts, radio stations, postage stamps, tourist activities, and the establishment of colonies. Chile's military has full control of the nation's Antarctic activities, and its Antarctic bases bear a striking resemblance to military installations (Parfit 1985, 270; personal observation, Teniente Marsh Base, Antarctic Peninsula 1986).

Under the Pinochet regime, Chile is especially sensitive to the presence of the Soviet Union and several of its allies, including Cuba, in Chile's claimed sector. There is much suspicion that Cuban and Soviet fishing boats in the South Pacific conduct espionage activities or logistically support armed subversive groups inside Chile. In the South Shetland islands along the west coast of the Antarctic Peninsula, there are relatively few ice-clear areas, and, as a result, many nations tend to concentrate their activities in just a few choice spots. On King George Island this has resulted in the placing of a Chilean, a Chinese (PRC), and a Soviet station (with Cubans) next to each other. While relations are correct between these nations, there is also the potential of an incident arising from this close contact (INACh 1984, 59–62).

Brazil

As indicated above, Brazil is the relative newcomer to the Antarctic arena of South American geopolitics. Its physical presence in the Antarctic Peninsula and its evolving Antarctic interests, are related to Brazil's perceived role as an emerging regional power with security and economic interests to defend in the broad region of the South Atlantic and the Antarctic continent. The first Brazilian Antarctic expedition was mounted in 1982, and acceptance as a full consultative member of the treaty system came the next year. The country's geopoliticians, however, were speaking of Brazil's Antarctic interest three decades before, and, indeed, it is to their credit that Brazil became directly involved in the continent.

Much of that involvement stems from the power of a single geopolitical concept, *defrontação* (frontage), which formed the basis for a possible Brazilian sector and which has undermined the strength of Argentine and

Chilean territorial claims, while at the same time appealing to Uruguay, Peru, and Ecuador as partners in a South American Antarctica.

The frontage theory was bitterly attacked by Argentine (and, to a lesser extent, Chilean) geopoliticians (Rodriguez 1974), who immediately saw it as a threat to their national geopolitical interests in the region. Although the frontage sector was never officially accepted as the position of the Brazilian government, it frequently appears in the Brazilian popular media and the specialized geopolitical publications. Further, it was advocated by a number of Brazilian geopolitical thinkers who played key roles in military governments from 1964 to 1985. Like the Pinochet regime in Chile, these governments were strongly influenced by geopolitical ideas, and the basic one was that Brazil had a geopolitical destiny of greatness as a potential world power. To reach that greatness, Brazil first had to become a credible regional power, and this meant projecting national influence in the Southern Cone of South American and beyond.

Its distance from Antarctica places Brazil at a disadvantage in terms of the potential rivals Argentina and Chile. Brazil had no serious historical claims in the South, except a possible tenuous one stemming from the Treaty of Tordesillas. Proximity arguments are not useful, although some authors claim that Brazil's climate is strongly influenced by Antarctica.

Lacking these arguments relied on by Argentina and Chile, Brazil is forced to fall back on the frontage theory and the idea that the Rio Treaty gives it a responsibility for the defense of the South Atlantic. There is also a geopolitical "coastline" argument that says that, since Brazil has the longest Atlantic coastline of any nation, it has a natural need and responsibility to secure the waters off that coastline, and this therefore takes Brazil to the far South Atlantic and to Antarctica. The NATO-SATO argument is also used, presenting the South Atlantic as a strategic vacuum the regional powers must fill, lest the superpowers move in.

Brazilian geopoliticians speak of a three-pronged national interest in Antarctica: security, ecology, and economics (Azambuja 1982, 275; Ibsen Gusmão Cámara, 1982, 22–23). The economic argument is based on current fishing activities and the possibility of a major energy source being developed in Antarctica. This would be of great interest for Brazil, which sorely lacks reliable energy resources to fuel industrial development. Even though the realists acknowledge that any such Antarctic energy would be a long time coming, they feel that Brazil must be well positioned for possible exploitation in the future (Leal 1974, 8–12; Moneta 1981, 52–53).

Brazilian Antarctic geopolitical analysts, like their Chilean counterparts, express concern over the presence of the Soviet Union and Warsaw Pact allies in "South American Antarctica." Recent articles have included maps showing how the Soviets and their allies have positioned their bases in Antarctica in a series of mutually supporting "strategic triangles" that could, in time of war, project power toward the South Atlantic, the Drake Passage, and other key interoceanic routes (de Castro 1984, 85–94).

Other Latin American Nations

The Antarctic geopolitical interests and activities of other Latin American nations lag far behind those of the three considered above, for reasons of distance, capabilities, and lack of solid and consistent geopolitical doctrine on Antarctic issues.

Uruguay has sent military personnel to Antarctica at the invitation of several nations with established bases and programs (mainly Chile and the United States). Uruguay set up its first base in 1984 and was accepted as a consultative party in the treaty system in 1985. Its Antarctic activities, though, are minimal and lack the extensive support system set up by Argentina, Chile, and Brazil. Uruguay's limited Antarctic activities are also a function of the geopolitical reality that it has always been a small buffer state caught between the two largest states of the Southern Cone, Brazil and Argentina. Perhaps because of a desire not to offend either neighbor, Uruguay has not stressed the frontage theory, but, rather, a pragmatic approach to presence in the region in collaboration with close neighbors. Much of Uruguay's South Atlantic and Antarctic interest has a historical base: the Spanish favored Montevideo over Buenos Aires as a port because of heavy silting problems on the Argentine bank of the Río de la Plata. The Spanish established their main *apostadero* (naval staging base) for the region in Montevideo, using it for many of the expeditions and sealing activities to the Malvinas/Falklands and the far South Atlantic (Fraga 1979). Uruguayan geopolitical thinking is also characterized by a strong integrationist streak, and this current shows up in arguments for a cooperative South American approach to the Antarctic area (Crawford 1982, 34–43; Vignali 1979).

Geopolitical consideration of Antarctica has recently received considerable impetus in Peru, and has stimulated official and popular interest to the point where it appears likely that Peru will soon have a permanent base in Antarctica (Mercado Jarrín 1984). This is especially noteworthy because geopolitical thinking in Peru took a different course than in Chile and Argentina, focusing on internal development issues much more than on external power projection. Like Brazil, Peru has little historical basis for a possible claim, and, as a result, Peru takes a strong geopolitical interest in the frontage theory, although its origin is attributed to Canadian, not Brazilian, sources. Frontage would give Peru a sector at the expense of Chile's claim. Peru's relations with its two South Pacific neighbors (Chile and Ecuador) have historically been difficult, and the press coverage of Antarctic interests has reflected these strains, especially when sensational headlines accuse neighbors of coveting "Peruvian rights in the Antarctic." In addition to frontage arguments, the Peruvians argue that the Rio Treaty and ecological concerns justify its interest in Antarctica. The ecological argument is of special significance because the cold Humboldt Current born in Antarctic waters has a considerable effect on Peru's climate and fishing industry.

Ecuador, at present, has only a minor interest in Antarctica, although the country was accepted as an acceding member of the Antarctic Treaty in 1987, and sent an expedition in the 1987-1988 summer season. Ecuador's interest, like Peru's, is based on the frontage theory, but in the Ecuadorean case the frontage would be derived from the Galápagos Islands, which lie some 600 miles out into the Pacific. The frontage stems from a 200-mile exclusive economic zone drawn around the islands. In terms of the geopolitical literature, the Antarctic theme appears sporadically in Ecuador, mainly in the army's geographic magazine and in some international relations texts (Villacres Moscoso 1984, 27–30).

Cuba has an active Antarctic presence, thanks to the Soviet Union, which regularly hosts a few Cuban scientists or medical personnel at its Antarctic bases. The Cuban flag flies below the Soviet's at the Bellingshausen station (King George Island), much to the irritation of the Chileans only a hundred yards away.

Several other Latin American nations have expressed interest in Antarctica, generally as part of the Third World current to make Antarctica's economic benefits "the heritage of all mankind" by denying sovereign or frontage approaches. These ideas have found their most favorable outlet in the General Assembly of the United Nations, which has included Antarctica as an agenda item since 1983. The Latin American nation most involved in this effort has been Antigua & Barbuda. Suriname, Mexico, and Bolivia have also expressed similar opinions (United Nations 1984).

POSSIBLE OUTCOMES

The wide range of interests and nations involved in the South American Antarctic Quadrant suggest that a number of possible outcomes may emerge as the psychologically significant year of 1991 is reached and passed. These outcomes can be grouped as follows:

Cooperative outcomes, with general agreement among the key players, would avoid serious confrontation. Given the vested interest of the superpowers and other member nations in the treaty system, the most likely outcome of all is probably a continuation of the present Antarctic Treaty System (ATS). This outcome would be enhanced if the ATS were gradually expanded to accommodate new members active in Antarctica, and, even more so, if an effective way of controlling mineral exploitation (the so-called minerals regime) could be implemented based on the June 1988 agreement signed in Wellington, New Zealand. But enlargement of the ATS cannot continue indefinitely or it will become a mini-United Nations, with all the discord and inefficiencies that this would entail. Two other cooperative outcomes, the internationalization ("legacy of all mankind") and the ecological ("world park"), seem unlikely, because of opposition by the ATS members, especially by the superpowers and the territorial ATS members.

Conflictive outcomes would result either from a breakdown of the ATS or from polarization among the various nations with strong Antarctic interests. It does not seem rational that any nation would deliberately want to bring down the ATS, but, as the 1982 Anglo-Argentine War illustrates, rationality may succumb to jingoism fed by highly nationalistic geopolitics. Under this scenario, one nation might take unilateral actions to make good its Antarctic claim or to establish control over a valuable resource. There may also be clashes between any two countries with Antarctic interests and presence, perhaps over a minor incident. Or there may be polarization between the ATS members and the excluded nations.

Mixed outcomes would involve cooperation between certain groups of countries which would then use their increased power to face outsiders. A South American (or Latin American) Quadrant Condominium would be an example: the nations in the Quadrant would cooperate in Antarctica and share any economic benefit, but would have to confront the excluded nations in the process. As this chapter has suggested, a strong current in South American geopolitical thinking argues for just such a cooperative approach; rarely does this type of thinking assess the problem of confronting the excluded nations, which would mean the United States and the Soviet Union, among others. Another mixed outcome might be the frontage-sector approach pitting Brazil and its three frontage partners against Argentina and Chile (and other non-hemispheric powers). There could also be cooperation between certain sets of nations (Argentina-Chile, Argentina-Brazil, Argentina-Chile-Brazil, Chile–Great Britain), a situation that would strengthen the Antarctic positions of the nations involved but would almost inevitably lead to confrontations with others.

Whatever the actual outcome, it seems clear that the South American nations will have a say in the process, especially if the outcome is influenced by incidents, activities, or resources in the area most likely to witness them: the Antarctic Peninsula, in which so many nations (including over a half-dozen Latin American ones) have an interest and presence. It also seems likely that the outcome, be it cooperative, conflictive, or mixed, will be influenced by trends of South American geopolitical thinking. There are indications that the old chauvinistic, aggressive, and nationalistic geopolitical rhetoric of the past has been giving way to a current of cooperative and integrative geopolitical thinking. But even this current may lead to confrontations if it unites the South American or Latin American nations, but pits them against outside nations who also have an interest in the political and economic development of the frozen continent.

chapter thirteen

Cooperation on Ice: The Potential of Collaboration in the Southern Cone

Margaret L. Clark

A most crucial factor to remember in discussing the issue of international politics in the Antarctic region is the longevity of its imprint. The Antarctic Treaty System (ATS) is often cited as an exemplar of international cooperation.

> The nuclear-free zone and the zone of peace created by the Antarctic Treaty serve as good examples for concluding similar agreements with respect to other regions of our planet and around it (Rybakov 1986, 36).

Recent international negotiations (such as the United Nations Conference on the Law of the Sea and the ongoing Antarctic minerals regime meetings) demonstrate that the human community is indeed attempting to organize the future of both frontier and resource management. Given the value of these regions, both geostrategically and resource wealthy, a state or region would be ill-advised to act in isolation.

The focus of this chapter is on the significance of Antarctic negotiations to the future of cooperation among the Southern Cone states. I will look at the issue facing north, that is, from the angle of Antarctic concerns, rather than from a Latin American perspective. As a result, the perspective is global and utilizes a lengthy time horizon. I will emphasize how the southern states use this opportunity of representation (or fail at such attempts) and will suggest strategic prescriptive behavior for use during the mineral negotiations of the "ABC" states. Mindful that these states do not always act in unison, this work begins with a brief introduction of the actors and their respective political positions on the issue of Antarctica.

The three states central to this discussion are Argentina, Brazil, and Chile (the "ABC" states). Both Argentina and Chile are original signatories to the Antarctic Treaty, which was ratified in 1961. These states are consultative members with full voting power. The two are also claimants to land area, including portions of the Antarctic Peninsula, both determining their claim of a sector based on two factors. The first of these is the geographic proximity of their states to the Antarctic continent (the Drake Passage is a mere 700 miles wide) and the related concept of continuity and contiguity. These reasons gain support from the histo-geological theory of the supercontinent Gondwana.[1] This geographic approach is reflected in Argentina's *Atlantártida* theory, which creates a geographical and political triangle of perceived influence and nationalistic pride. Chile's geopolitical position is a similar one, affected by both the transoceanic passages (Magellan and Drake) and the close geographic proximity to Antarctica (Child 1985, 140–141).

The second factor that influences the claims of the two states is the 1493 Bull Inter Caetera of Pope Alexander VI. This aspect comes under scrutiny by some contemporary scholars who contend that it "applies only to lands and islands discovered or to be discovered by one of the two countries" (Auburn 1982, 49).

There are no historical records to indicate that this stipulation was fulfilled. However, the papal bull position is currently supported by Argentina and Chile, which cite *uti possidetis*, a general practice of arbitration for intra-American laws.[2] It is often attacked by states outside the region who question the accuracy of the information used to render early state border decisions. According to this *uti possidetis* perspective,

> the South American sectors were not res nullius at the time when Britain's various acts of sovereignty were carried out (Auburn 1982, 50).

Consequently, Britain's claim to an Antarctic sector, which overlaps the claims of the two Latin American states, is void. Britain counters that *uti possidetis* is a regional custom and is therefore subordinate to international law, so that the sector in question was *res nullius* until British arrival.

Despite their individual commitments to state goals, including securing an Antarctic claim, both Argentina and Chile have "repeatedly put forward a united front against Britain" (Auburn 1982, 55). Such cooperation was manifested in the 1948 Vergara–La Roze Declaration between the two states. It recognized the external limits of their claims and also formed an agreement to act jointly in Antarctic affairs (Maquieira 1986, 51).

The third Latin American state with consistent Antarctic involvement is Brazil. Although the state has participated in Antarctic research projects since the International Geophysical Year (1957–1958), it did not adhere to the ATS until 1975 and did not acquire consultative status until 1983. Official Brazilian reasons cited for such delays include the perceived priority of internal or domestic development. The official state policy remained opposed to territorial claims on both historical and future-oriented grounds:

It seemed to be much more advisable to maintain our freedom of action. . . . A territorial claim . . . was not advisable, and it would restrict rather than enlarge the opportunities offered by Antarctica (Guimaraes 1986, 339).

Brazil's more recent contribution to the Antarctic debate has produced some controversy. The defrontação, or frontage, theory is based on a similarly ascribed Arctic sector policy. According to this approach, each South American state with unobstructed meridians to the South Pole would be entitled to an Antarctic sector (Child 1985, 134; Pittman 1986, 50–52). This would create a "South American Antarctic Sector" that would effectively deny territorial rights to other states, especially to Great Britain. Although strongly supported by Uruguay, Peru, and Ecuador, this position is opposed by Great Britain, Argentina, and Chile. As a result of such hypothetical gerrymandering, Brazil would acquire the largest regional sector at the expense of both Argentina and Chile. These geostrategic alterations would be likely to exacerbate the present regional rivalries. Such a theory is unlikely to reach fruition. This is partially because such dramatic changes would be met with substantial resistance from other ATS members.[3]

These three South American states are varied in their positions regarding Antarctica. Argentina continues to teach students about Atlantártida, Chile has similar territorial theories, and Brazil presses for the frontage approach to the region. However, their positions also reflect the tendency of states to protect their individual goals and to preserve their national pride. Before considering the destructive aspect of these individual agendas, it is valuable to consider the wider regional scenario.

Scholars of this region are familiar with the various Antarctic links to other local conflicts. One such example is the Southern Andean conflict between Argentina and Chile. This struggle involves the Beagle Channel Islands and the corresponding political condition of the Southern Atlantic. A more recently addressed issue relates to the logistic and strategic importance of the Malvinas/Falkland Islands, which also reflects the geostrategic importance of the entire region.

[Antarctica is the] land mass between the southern portions of three major oceans . . . and [is therefore] linked to regional and superpower attempts to project power in these areas (Child 1985, 133).

From an even broader perspective, the international arena, the issue of Antarctica creates interesting challenges to the Southern Cone region. One of these is the concept of "internationalization" of the Antarctic region, which is opposed by both Argentina and Chile (Child 1985, 140). The term "internationalization" as used here does not refer to the expansion of either the ATS membership or the responsibilities under that agreement. Rather, the term defines the transfer of management of the region (and its resources) from the ATS members to a more universal body, such as the United Nations.

Such a concept is not new. In 1948 the United States proposed an "international administration" of Antarctica by seven claimant states, with the United States acquiring the "unclaimed" sector. This would have created an eight-power trusteeship under the United Nations (Maquieira 1986, 52). While this would have resolved the individual claimant issues, it also could have barred the Soviet Union from any future participation. (This U.S.-influenced proposal was, most likely, in response to the Berlin Blockade, which also occurred in 1948.) Such an elite trusteeship met with opposition from various states. Consequently, the United States shifted its position toward the management of the region by a condominium of states. Although supported by Great Britain, the proposition was rejected by both Argentina and Chile. The latter state offered a counterproposal, the Escudero Declaration, which involved a five-year moratorium on territorial claims.

In some cases, links have also been forged with organizations outside the United Nations system. These types of relations were considered valuable during the early stages of the treaty's negotiations and are therefore formalized in both Article 3 [Section 2] and Article 11 of the Antarctic Treaty. Although one could argue that such links create a degree of internationalism, the power of the treaty members is not thereby jeopardized. The expertise of these organizations (for example, in the fields of meteorology and communications) serves two useful purposes. First, the increased knowledge enhances the various contributions of Antarctic research. Second, the relations between these organizations and the ATS members expand the range of states that can participate in and benefit from Antarctic research. These links and the continued studies undertaken by the United Nations indicate that the future use (or non-use) of the region's resources will not be determined without the representation of diverse interests.

These factors underscored the contemporary changes in the international community. Specifically this involved

> the emergence of newly independent nations [during] the 1960s [that] have brought demands for change in existing international legal rules and regulations (Qasim and Raja 1986, 368).

ATS is considered by some to be an example of an elite form of decisionmaking behavior. Some actions have been undertaken in an attempt to rectify this condition. For example, the dissemination of Antarctic information has been expanded via two channels: the office of the United Nations secretary general and the Scientific Committee on Antarctic Research (SCAR). Two new research bases have also been established, India's "Southern Ganges" in 1983 and the People's Republic of China's "Great Wall" in 1985. Some states with greater research capabilities (such as equipment and field stations) have continued to include, within their field parties, scientists from less advantaged states. As a result, the list of states actively participating in projects in the region (and therefore directly benefiting from the research) has expanded.[4]

Such expansionism has produced changes both in the status of some ATS members and the inclusion of new ones. In 1987-1988, five states became new contracting members: Greece (January 8, 1987), the Democratic Republic of Korea (January 21, 1987) Ecuador (September 15, 1987), and Canada (May 4, 1988). In 1983, Brazil and India both acquired consultative status, and the Republic of Korea plans to establish a base during the 1987/88 field season, thus qualifying for consultative status. A dramatic change occurred in 1983 when it was determined that

> all parties to the Antarctic Treaty may attend consultative meetings . . . and future meetings of mineral negotiations as observers. . . . . [As a result of this behavior, states can accede to the ATS] at no cost . . . [and can] participate in and influence decisions taken in these meetings (Woolcott 1986, 383).

The resulting involvement of the so-called developing world, coupled with the expanded dissemination of information led some parties to conclude that

> the distinction between [the] Antarctic Treaty consultative parties and non-consultative parties is likely to become less marked in the future (Woolcott 1986, 383).

This recent growth of the developing world's participation also creates the impression that those parties are satisfied with the composition of the Antarctic decisionmaking institutions. But the fact remains that these institutions retain a two-tier decisionmaking process in which one tier has greater influence. This arrangement means that the developing states rely heavily upon those higher-tier members who represent similar Second and Third World views. Argentina, Chile, and Brazil are perceived by many in the developing world to represent not only Latin American views, but also Antarctic concerns of the Second and Third Worlds more generally. Consequently, these three states are in an enviable position of representing a fair percentage of the world's population, as well as an entire region.

Unfortunately, the diversity of the three states' individual goals appears to undermine any integration efforts:

> Currents of geopolitical thinking that stress integration and harmony, and that would presumably tend to make South American conflict less likely, are not the prevailing ones (Child 1985, 33).

Child states that nationalism and a suspicion of neighboring states is the dominant condition in the region, especially regarding questions of resources and territorial claims. Certainly such topics are the primary focus of Antarctic politics:

The most significant potential endowments . . . are those within the Dufek complex and [the oil resources of] the Weddell Sea. Both of these areas lie within the parts of Antarctica claimed by Chile, Argentina and the United Kingdom (de Wit 1985, 68).

Ideally, the Latin American states should acknowledge that their political position on the issue of the Antarctic has substantial long-term advantages. If they perform well as leaders in this issue, these states may gain future positions of leadership in the developing world. Likewise, a strong, cohesive Latin American stand on the issue of Antarctica will make a positive impression in the larger international arena.

This point is stressed by L. F. Macedo de Soares Guimaraes of the Brazilian Ministry of External Relations, in discussing Brazil's new Antarctic position. During the mineral negotiations, it became clear that an "accommodation" approach to decisionmaking was necessary not only between the territorial claimants and the nonclaimants, but also between the industrialized and the developing states. This meant that the needs and concerns of both developed and developing participants had to be acknowledged and considered. This revelation was a reaction (at least in part) to Brazil's new consulative status and continued allegiance to the interests and needs of the developing world. Although Brazil's primary concern is to represent its own specific domestic needs, it also

> necessarily acts as a developing country. . . . [Brazil] will contribute to the defense of the interests of the developing countries. . . . The [Antarctic] community has to change its ways in order to adapt to the new member. . . . New forms of cooperation must be established (Guimaraes 1986, 343).

In short, Brazil's policy objective has been to fulfill its internal needs and goals without abandoning the larger priorities of the developing states.

Chile and Argentina, despite their own conflicts and rivalries, should seriously consider the long-range benefits of following Brazil's example. These three states are in stable leadership positions, having unquestionable voting rights within the Antarctic Treaty System. If these states maintain fractious relations, however, they fail the developing world as well as themselves. In such a case, resource decisions (and other related decisions) could be made so as to benefit the industrial states at the expense of the developing states. Therefore, a key factor is to identify some methods by which the ABC states can establish a feasible, cooperative relationship.

One form of ad hoc cooperation that could be pursued by the ABC states is "functionalism," described by Mitrany. A primary goal of functionalist strategy is to

> call forth to the highest degree possible the active forces and oppor- tunities for cooperation, while touching as little as possible the

latent or active points of difference and opposition (Mitrany 1943, 58).

Cooperative actions can generate a temporary "sympathetic federation." Such mechanisms would organize activities around the locus of common interests and tasks, thus demonstrating a readiness to cooperate for advancing those commonalities (Mitrany 1943, 55). This does not signal the necessity of a separate organization for each issue, but rather the necessity of practiced cooperation.

Such cooperation may appear to be beyond the capabilities of Chile, Argentina, and Brazil, given their diverse political opinions, historical animosities, and national prides. But, Axelrod and Keohane state that cooperation is not necessarily synonymous with harmony, rather

> cooperation can only take place in situations that contain a mixture of conflicting and complementary interests. . . . Cooperation occurs when actors adjust their behavior to the actual or anticipated preferences of others (Axelrod and Keohane 1986, 226).

In the issue of Antarctic politics these South American states have already committed themselves to active participation. As previously mentioned, the pending question for the ABC states is to define what form of "active participation" they will create. Their potential complementary interests are threefold. The first involves increasing their participation in international negotiations, especially those that deal with the future distribution of resources. The second is to benefit from both the scientific and market values of actual and potential Antarctic resources. The third complementary interest lies in the defense of the specific interests of the developing world. This final interest, if pursued consistently and effectively, would have long-term benefits for the ABC states. Not only would these actors gain status among the other developing states, but they would also improve their position among the more industrialized states because of their increased credibility as Latin American negotiators. The three states should also take precautions in order to preserve the priority of these interests and not to succumb to "alliances" with industrial states that may later threaten those interests. Such apparently benign proposals are defined by Oye as

> offers [in return for compensation] to refrain from acting in what would otherwise be [one's] own best interests (Oye 1986, 240).

This refers to those propositions made to some states to provide incentives for behaviors counter to their best interests. The production of cash crops is one example of such a proposal. A state agrees to devote a substantial amount of arable land to grow a specific crop. In return, the other state will provide a ready market and preferential trade agreements. While this may be mutually advantageous, over time it will convert the producer state into one with a dependent economic base.

Such agreements between parties can conceal a compensation that, when fulfillment is requested, can be burdensome to one of the actors involved. As a result, parties can find themselves embroiled in long-term commitments that are no longer beneficial. These situations can be particularly tempting when the initial offer of assistance is perceived as a necessity for internal or international development.

One method by which actors can attempt to prevent this entanglement is through collective action. Argentina, Chile, and Brazil could form such a group utilizing a "sympathetic-federation" approach. Initially, the collectivity would be negotiating in an environment dominated by piecemeal, or "salami," tactics that tend to benefit the industrialized states. These tactics would rely on the historically suspicious relations among the three states, as well as their strong nationalistic nature. Overcoming these assumptions held by the industrialized states (at least, as those beliefs affect Antarctic positions) would be a worthwhile effort for the Southern Cone states. Before concluding that such an ad hoc cooperative is forever illusive, a further investigation of Axelrod and Keohane's definition and development of "cooperation" may prove illuminating.

Three factors affect an actor's propensity to cooperate: (1) the mutuality of its interests with those of other actors, (2) the size of the shadow of the future, and (3) the number of actors involved and the content of their inter-actions (Axelrod and Keohane 1986, 229). In the case of Antarctica, the first factor has already been discussed. All three states have vested interests in the future of the region and its resources. Consequently, one mutual interest is to improve their negotiating reputation on this issue. A potential problem could exist in their contradictory perceptions of events and responses. The importance of accurate information regarding events and the positions of the other actors needs to be stressed (Jervis 1982). The suspicions among the Southern Cone states should not be permitted to interfere with the transmission of information and other communications. Pursuit of open communication would reduce the likelihood of a state's actively responding to inaccurate information and thereby causing political upheaval. It would also provide the states with an initial, nonpolitical cooperative task, providing access to correct information.

The second factor, the "shadow of the future," is important because it contributes to the determination of the future value of benefits resulting from present actions or promises. Axelrod and Keohane describe such a shadow as existing when

> The more future payoffs are valued relative to current payoffs, the less the incentive to defect today—since the other side is likely to retaliate tomorrow (1986, 232).

The "shadow of the future," then, provides actors with perceptions and expectations about the future role or impact of their present behavior. This information can then be used as self-inducement to achieve desired behavior. According to Axelrod and Keohane (1986), the "shadow of the future" involves four components: (1) the length of time horizons, (2) the regularity of future stakes, (3) the reliability of information regarding the actions of others (as discussed in the above paragraph), and (4) the mechanisms for feedback about such external activities.

The first component (time horizons) involves the continued debate over the existence of marketable mineral and hydrocarbon resources. The necessary technology for safe and successful extraction has yet to be developed. Therefore, the current Antarctic resource negotiations are being conducted in substantial advance of the implementation of their results. These long time horizons, which are inherent in Antarctic negotiations, are valuable to the parties involved. The developing states acquire negotiating experience that reduces the possibility of their exclusion from future resource negotiations. Similarly, the industrial states want to protect and improve their positions of technological advantage. Nonstate actors (such as international organizations) also prefer long time horizons for two primary reasons. The first is that time provides these groups with more opportunities to gain admission to the negotiations (although their presence is usually limited to that of observers). The second is that time can be constructively utilized to demonstrate their capabilities and knowledge to other groups and states, thereby improving their credibility.

The second factor which affects Axelrod's and Keohane's "shadow of the future" is the regularity of future stakes. It is reasonable to assume that the needs of the industrial states will continue to influence strongly (if not to dictate) the market for minerals and hydrocarbons. Since the economies of these states rely heavily upon such resources, the demand for them is unlikely to diminish. Consequently, the market value of the resources (not necessarily the market price) is relatively stable. As long as the existence of mineral and hydrocarbon resources in the Antarctic is a possibility, there will be some interest in exploration and extraction. For these reasons, the future stake of Antarctic involvement appears to be a stable and worthy investment for the creation of a sympathetic federation among Argentina, Chile, and Brazil.

The last factor affecting an actor's ability to cooperate is the total number of actors involved and the nature of their interrelationships. In most cases, a smaller number of actors converts into a more manageable situation. The balance between participants in such structures depends upon the existence of reciprocity among participants in order to induce cooperation. These three factors are not beyond the ability of the ABC states. Pittman (1986, 55) refers to one tenet of Argentine geopolitical theory that highlights such confidence in cooperation:

> the South American continent should be integrated into a single monolithic economic and power bloc in order to be able to deal

internationally with the superpowers and the EEC on a more equitable basis.

This chapter has focused on reducing the myth that cooperation is synonymous with harmony. It has instead discussed the components of feasible cooperation and how these components can be acquired by the Southern Cone states. As this chapter has shown, Argentina, Chile, and Brazil have certain complementary interests. Strategies of participation in international negotiations, acquisition of benefits from Antarctic research, and leadership among the developing states could be of political significance to all three parties. A "sympathetic federation" could thus emerge to facilitate achievement of mutual goals and those of the larger developed world.

Such an endeavor would not require that Argentina, Brazil, or Chile relinquish its sovereignty, or even a significant portion of it. Rather, such measures would aid in the dismantling of interstate barriers that too frequently inhibit even the consideration of cooperation. The resulting federation would necessitate cooperative efforts by all the parties involved. I have discussed the elements that indicate an actor's propensity to cooperate (a mutuality of interests, a shadow of the future, and the number of participants and their interrelations), and have assessed these components according to the specific capabilities and needs of the ABC states. I have concluded that the tradition of equating harmony and cooperation is highly detrimental. Such sentiment justifies the failure to attempt cooperative actions with those actors outside the established "friendly parameters." Consequently, actors do not challenge the status quo perceptions of future interactions. Before any interaction is even attempted, its failure is seen as a foregone conclusion.

While it is true that Argentina, Brazil, and Chile may never practice full cohesion in all international negotiations, ad hoc federations on specific topics are feasible. Areas likely to be productively discussed are those in which all participants have a vested interest. Antarctica is such a topic. The potential existence of marketable resources is a factor, but the future direction of international negotiations and law is a more immediate and recognizable consideration. The leadership of the ABC states in the Antarctic Treaty System is an enviable one. It would be a great loss for these states not to take full advantage of their positions, both within the Southern Cone region and within the larger international community.

NOTES

1. This ancient continent consisted of what is today Antarctica, Australia, Africa, and South America. The gradual breaking up of Gondwana was the result of continental drifting. It was not until the mid-Permian period, some 250 million years ago, that Antarctica began to slide southward (Shapley 1985, 125–127).

2. In international law *uti possidetis* is a phrase used to signify that parties to a treaty are to retain possession of what they have acquired by force during a war (Wheaton's International Law, 627).

3. In his contribution to the 1986 Beardmore Conference proceedings, Guimaraes does not mention this frontage theory. It is unclear whether Brazil has abandoned this theory, or whether Guimaraes simply chose not to include it in his presentation.

4. The Republic of Korea became a member of the CCAMLR (the Convention on the Conservation of Antarctica Marine Living Resources, a component of the ATS) commission on November 19, 1985. By this action, the ROK was able to demonstrate its sincere interest and involvement in the region. It subsequently became a contracting member on November 28, 1986.

chapter fourteen

The Strategic Importance of the South Atlantic

Carlos de Meira Mattos

STRATEGIC CONSIDERATIONS REGARDING MARITIME POWER

The military or strategic value of a naval position depends upon its situation, upon its strength, and upon its resources. Of the three, the first is of most consequence, because it results from the nature of things; whereas the two latter, when deficient, can be supplied artificially, in whole or in part. Fortifications remedy the weaknesses of a position, foresight accumulates beforehand the resources which nature does not yield on the spot; but it is not within the power of man to change the geographical situation of a point which lies outside the limit of strategic effect.

—Admiral Alfred Thayer Mahan
The Interest of America in Sea Power, Present and Future (1897)

The accomplished studies of Professor Lewis Tambs of Arizona State University conclude that a major part of global maritime commerce passes through fourteen choke points:

- Five interior seas (the Mediterranean, the North, the Norwegian, the Caribbean, and the South China)
- Two interoceanic canals (the Suez and the Panama)
- Seven critical maritime passages (the Mozambique, the Horn of Africa, Gibraltar, the Magellan Strait, the Cape of Good Hope, and the Straits of Sri Lanka and Malacca).

These fourteen choke points have been an arena of conflict since the sixteenth century, when maritime expansion began among the European powers.

Translated by Philip Kelly

When we consider the wide extent of the Atlantic Basin, including the seas and the connecting routes to this ocean, we find that nine of these fourteen points are associated with this basin:

- In the South Atlantic: the Strait of Magellan, the Cape of Good Hope, and the Mozambique Passage
- In the North Atlantic: the Straits of Gibraltar and Panama, the Caribbean, the Mediterranean, the North, and the Norwegian Seas

This predominance of spaces that control maritime commerce emphasizes the importance of the Atlantic Ocean in the context of world maritime strategy.

Before we consider a particular strategy for the South Atlantic, we must make clear our viewpoint that the Atlantic represents a great geostrategic unity and should not be thought of in terms of isolated strategies within the general area. In case of a world, armed conflagration, the strategies of the North Atlantic and of the South Atlantic complement each other and ought to support each other reciprocally.

THE SOUTH ATLANTIC IN THE CONTEXT OF WORLD POWER

The strategic importance of a maritime area should be defined according to the potentialities of its use:

- As a route of commercial transportation or of military force
- As an area for projection of military power over territory
- As a source of resources.

The employment of maritime power of a country or group of countries should restrict to its advantage the use of maritime areas in each of the three aspects indicated above, and, simultaneously, deny these to the enemy. Keeping in mind this basic doctrinal concept of naval warfare, we propose to analyze the theme already introduced, the strategic importance of the South Atlantic.

There is no precise or widely accepted definition of the geographic boundaries of the South Atlantic. Some authorities define its waters as being south of the Equator. The criterion established by the Atlantic Treaty, which considers the North Atlantic as the maritime zone north of the Tropic of Cancer, identifies South Atlantic waters to be south of this geodesic line. A more adequate and natural geographic definition, and one that we adopt for this presentation, is to consider only the mass of oceanic water to the south of a line from the Brazilian northeast salient and the northwest African prominence of the Natal-Dakar strait.

All of the Brazilian coast from north of Natal, then, in effect faces toward the North Atlantic. In truth, Natal is equally distant from Washington and Bordeaux, as are these cities from each other.

Map 14.1   South Atlantic Distances

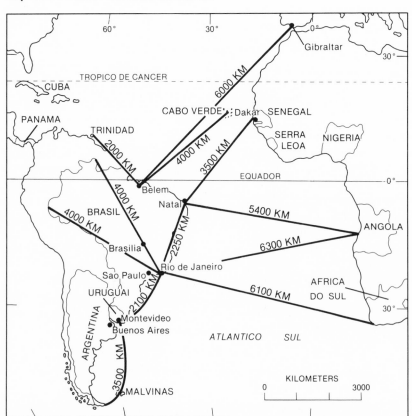

The South Atlantic has three accesses, areas of straits which connect oceans: the Natal-Dakar strait, the passages to the Indian Ocean, and the passages to the Pacific in the extreme south of South America (Magellan, Beagle, and Drake passages). The strategic limitations and vulnerabilities posed by the Panama Canal can only be resolved by use of the inter-oceanic passages located in the extreme south of the continent, which would be more secure in case of world military conflict.

We must consider a South Atlantic strategy for the year 2000, when technology will almost be able to exert supremacy over various limitations in frozen regions, changing the importance of the continental position of Antarctica in relation to the nearness of its northern extremity to South America (the Drake Passage), its position facing the region of the Cape of Good Hope, and its mineral wealth that technology alone may open.

Nations of the southern Atlantic are forced to coexist with military, political, and ideological competition between two groups of nations—the Western world and the Socialist nations, these militarily within the Soviet nucleus. We consider this a valid dichotomy, while still recognizing the existence of numerous dissenters, divergencies, and even internal antagonisms in each of the two groups. North-South rivalries (developed and underdeveloped states) and China against the Soviet Union are the most glaring cases of this.

From a strategic point of view, as long as the Socialist nations are grouped only in land masses that are in large part self-sufficient in natural resources and in primary goods, the Western world will disperse itself on continents, islands, and peninsulas, and they will be completely dependent on existing resources from distant regions that will require maritime transport as their only valid supply option. This dependence on maritime transport and the growing demand for natural resources and markets, so characteristic of contemporary civilization, has caused a remarkable growth of international commerce and of the number of navies whose protection this commerce requires.

A map indicating all of the navies navigating the seas of the world (in 1969, attaining a total of nearly 21,000 units, with an estimate of 40,000 in 1990) shows that the commerce of the South Atlantic tends to be concentrated near the coasts of South America and of Africa and between the two salients of these continents. This contrast, of great significance for defense purposes, comes about in large part because many of the lines of maritime traffic in the North Atlantic originate and terminate in the coasts of this same ocean, whereas in the South Atlantic, they originate and/or terminate in other oceans, notably in the North Atlantic and in the Indian Ocean. Particularly important is the intensive traffic encompassing Africa, which moves around the continent's coast, with, for the most part, immense quantities of Middle East petroleum imported for the Americas and for Europe. For this reason, the Cape of Good Hope, today occupying third place in total global capacity shipped in maritime traffic, will probably occupy second place by the beginning of 1990. Today, the English Channel is the first and Gibraltar the second in such transit.

Turning to Brazil and its maritime interest, we are able to describe the country as equivalent to an insular region because more than ninety-five percent of its total foreign commerce is transported by sea. This commerce is still growing, with cargo increases substantial, passing from 33 million tons in 1963 to 300 million in 1986. All of this immense quantity of merchandise is dispersed throughout the world in many forms or supplied in distant areas among many populations, but it is important to stress that more than eighty percent of the total is transported across naval routes passing near the African coast. Hardly twenty percent corresponds to inter-American maritime traffic.

These facts permit us to argue that, in the case of Brazil, the strategic routes of the Atlantic Ocean are principally those which are directed to Europe and to southern Africa (with a destination to the Middle East or to

Japan), and those which follow the American coasts to the River Plate and to North America. What stands out in economic and military significance is the oil route to the Middle East.

Besides maritime traffic, another use of these oceans is the extraction of natural resources. Petroleum is exploited principally near the coasts of Brazil and Argentina, and in Nigeria, Gabon, and Angola. The off-shore oil production of Brazil, already substantial, is growing and appears to be strategically important because it is not dependent on the extremely long and vulnerable lines of maritime transport. The African production, not itself marine but coastal, even now is significant (the major producer, Nigeria, processes nearly two million barrels each day), but it is still not comparable to the Middle East, and nothing indicates that it is going to be. Just the same, in a war emergency, it would be of enormous advantage and would be avidly disputed if Middle Eastern production became blocked.

Fishing is concentrated principally on the continental shelves of South America and of Africa, and in the ocean areas near Antarctica. One of the countries most dedicated to the fishing industry in the South Atlantic is the Soviet Union. It has been active in Antarctica for some time, and, to make matters worse, with each year the Soviet Union operates more intensely along the coasts of certain African nations. Despite its great contribution toward satisfying world food needs, however, fishing in the South Atlantic is not of sufficient importance to create problems of a strategic nature.

We must examine collective security in the Atlantic area in terms of two specific treaties (thus excluding the vague defense commitments and the seldom stable expressions of Articles 42 and 43 of the United Nations Charter): the Inter-American Treaty of Reciprocal Assistance (the Rio Pact) and the North Atlantic Treaty, the latter with its respective organization, NATO.

The oceanic area included under the Rio Pact is restricted to the western part of the Atlantic. The area of interest for NATO embraces only the North Atlantic above the Tropic of Cancer. The main part of the South Atlantic, therefore, is not covered formally within these two collective-security treaties. Yet, any East-West global military crisis would involve, decisively, the maritime supply routes of the South Atlantic, obligating the belligerent northern nations to intervene with their naval forces into this ocean. This hypothesis became clearer after the Malvinas conflict, when England, a NATO power, moved to occupy militarily this extreme southern archipelago of the South Atlantic. We must also recognize that the Rio Pact, because of its character, should not be considered a military alliance. The nonbinding nature of armed participation in case of aggression on the part of its member states, and the nonexistence of an organized permanent military structure during peacetime (such as exists under NATO) in large part restricts the effectiveness of the pact for collective-security ends in the case of armed conflict.

With the Antarctic Treaty, in force since 1961 and subject to possible review in 1991, member states (among these, Argentina, Brazil, the United States, and the Soviet Union) have agreed to utilize that continent only for peaceful ends. The involvement of the Antarctic continent in future armed conflict is consequently excluded.

We also cite the Tlatelolco Treaty, which forbids the use of nuclear weapons in Latin America and in adjacent maritime areas. This treaty, however, has not yet had its validity fully substantiated, for this legality depends on the ratifications of some of the participating states. We mention finally that the General Assembly of the United Nations approved a resolution declaring "the South Atlantic as a Zone of Peace and Cooperation." This resolution was originated by a Brazilian proposal.

STRATEGIC CONSIDERATIONS

The complex strategic panorama of present times is totally determined by a new factor in the history of warfare—the nuclear equilibrium and the reciprocal fear of such war between the two nuclear superpowers. This situation of nuclear parity makes a military confrontation between the Soviet Union and the United States highly improbable because the risks of mutual destruction would be unacceptable for whatever political objectives might be sought. Today, everything takes place as if there were, in fact, a desire to perpetuate the nuclear impasse, since the only guarantee for durable peace depends on the superpowers being strongly armed. The supremely powerful strategic nuclear arsenal works through intimidation; if, some day, it is actually used, it would fail in its final usefulness. For this reason, it is indispensable that these nuclear forces be complemented by conventional armaments of less power yet of controlled utility in the political game of calculated threats, demonstrations of force, wars of revolution, and limited conflicts.

The improbability and unacceptability that the nuclear equilibrium will turn into a direct military conflict among the nuclear nations paradoxically has created, by these very facts, a favorable atmosphere for the appearance of numerous small wars and crises of limited scope, with many of these connected to the political or strategic interests of the Great Powers. Likewise, there is at present a curious situation of near stability between the two superpowers at the same time as there are growing levels of violence and proliferation of localized military conflicts in more isolated geographic areas. The greatest portion of these conflicts is supported by and openly supplied from the Soviet Union, which feels the conflicts are useful to the achievement of Socialist political objectives. The involvement of Cuba in this turmoil, with the massive logistical support of the Soviet Union, is a new factor of great importance. The erection of Soviet footholds on the

African continent will enable an extension of Socialist influence to other regions in the future.

In this confused world and regional scene, where interpretation becomes progressively more difficult, an irrefutable fact stands out clearly—a growing presence of Soviet military power in regions that only a few years ago were totally isolated from such operations. While these favorable ideological and political footholds provide new areas for logistical support, Soviet naval-air power will also have an impact, backing new thrusts and potential threats to the maritime interests of Western nations.

The maintenance of maritime traffic flow is the principal interest of the West in the South Atlantic (and in the Indian Ocean as well). This interest derives from the unrestrainable need for petroleum brought on by the wastefulness of technological societies. Enemy control of vital strategic choke points on the African continent contributes particularly well to dangerous political blackmail in time of peace, and, in the event of an escalation to military conflict, to strangulation of the Western economy. The lessons of the oil embargo that followed the 1973 conflict should not be forgotten.

The rise of major Soviet influence in areas on the African continent and in the Middle East presents a novel coincidence with the strategic interest of the West for dominating maritime traffic and for controlling the sources of natural resources, especially that of petroleum.

We must recognize that, at the moment, the presence of Soviet naval-air forces in the South Atlantic, although practically constant in strength, still is not very significant in a numerical sense. Nevertheless, considering Soviet expansionist tendencies in the past decade, we should not be surprised in the near future if this new force would become an important military factor within the region. Operating from African bases, Soviet submarines and aircraft, assisted by surface forces and sophisticated processes of naval reconnaissance (including satellites), would be able to threaten all traffic in the vicinity of Africa, extending this power also to American coasts and to a substantial part of the North Atlantic.

For the West, in the present or previous state of affairs, the defense of the South Atlantic means basically maintaining the continuity of maritime traffic and preserving the operational capacity of its air and naval forces. For Brazil, beyond these common purposes, it must also protect its off-shore petroleum fields and protect against direct attacks on its territory and coastal installations.

We ought likewise to remember that Brazil occupies an additional strategic position in the Atlantic, that of projecting itself into the ocean at its narrowest part (the Atlantic Narrows, which Roosevelt called the Natal-Dakar strategic bridge) and, furthermore, having its immense coast turned toward the South Atlantic as much as toward the North Atlantic. If the African coasts come under the influence of enemy forces, the strategic importance of the

Brazilian geographic position will grow, as happened during World War II; Brazil will be able to fulfill this role again, to involve itself in conflict that it did not create. Consequently, in case of war or of international tensions, Brazilian neutrality would remain threatened in proportion to the level of Eastern nations' presence.

The interoceanic passages to the extreme south of the continent (the Magellan, Beagle, and Drake), as well as the islands and archipelagoes that serve as supports to navigation in this strategic region, will come to play a role of major importance in maritime strategy in an unforeseeable third world war. This southern passage would be the only interoceanic route open to navigation from any port, since the Panama Canal could be blocked in the first hour (and presents difficulties and limitations of navigation even in normal times).

As a final strategic consideration, we must remember that the Soviet navy, although appearing as a growing threat to the West, also confronts its own difficulties—inevitable geostrategic limitations stand out, in the form of having distant metropolitan bases and of the wholesale aging of a good part of its submarine fleet. In view of the recent tendency of the U.S. Navy to return to its old vigor and of similar occurrences in other Western navies contributing their share toward defense, the equation of the balance of forces should henceforth evolve in directions more favorable to the West, neutralizing to a certain point the previous political and strategic gains of the Soviet Union in Africa—so long as Western governments recognize these new conditions, and adequate political actions continue the improvements.

STRATEGIC OPTIONS FOR THE COUNTRIES
OF THE SOUTH ATLANTIC

From the point of view of global strategy, the South Atlantic occupies a secondary position in comparison to the North Atlantic. I hold no doubts that the center of the world military crisis is concentrated in the North, nor would I debate that the importance of commercial interests in the area are underestimated. Universally, the North Atlantic is more important. The South Atlantic today figures within the Western military strategy particularly as a supply route for the principal volume of oil coming to the North from the Middle East. In the case of world conflagration, the interoceanic passages of the extreme south of the South American continent (the Magellan, Beagle, and Drake) will grow substantially in importance.

In the hypothesis of global conflict, confronted by the belligerent antagonism between East and West blocs, the countries of the South Atlantic would inevitably be involved. They must choose between a difficult neutrality being respected and being participants within one of the leading blocs of the two superpowers. Neutrality of these countries might be better tolerated if they were not located in such a vital strategic position for the principal Great Power contenders.

The choice of becoming allied to one of the blocs will be a crucial decision for those countries that occupy important geostrategic positions. In the South Atlantic, the following positions stand out in this order of strategic value: the region of the Cape of Good Hope, the two salients of the Atlantic Narrows (on the one side, northeastern Brazil, and, on the other, Senegal, Gambia, Guinea-Bissau), and the region of straits in the extreme south of the American continent (Magellan, Beagle, and Drake). Armed conflict between the two great blocs would intensify pressure for gaining dominance of these positions, and such conflict could become irresistible.

This harsh reality requires that the countries occupying the geostrategic regions emphasized above consider clearly their international policies within a hypothesis of world conflagration and the consequences they would face. It is preferable not to have illusions. The natural line—historically, ideologically, geographically, and politically—of those South Atlantic countries whose territories would be involved in this major war is much closer to the Western democratic-Christian system than to the Eastern socialist-totalitarian system. This choice, although reluctantly made, should not become a position that contradicts this heritage.

chapter fifteen

The Geopolitics of the Falklands/Malvinas and the South Atlantic: British and Argentine Perceptions, Misperceptions, and Rivalries

Leslie W. Hepple

The dispute over the bleak islands in the South Atlantic known to the British as the *Falkland Islands* and to the Argentinians as *Las Islas Malvinas* has a long history and a very substantial literature. The war of 1982, with discussions of its causes, the campaign, and its consequences, has generated hundreds of books and papers. This chapter sets out to explore one particular aspect that has remained largely ignored, especially in the English-language literature. This is the geopolitical perspective. Geopolitical analysis has had an influential role in Argentina, particularly (but not only) in military circles. The argument of this chapter is that British and Argentine commentators and policymakers have had very different geopolitical perceptions, that this has led them to misperceive and misunderstand each other's motivations, and that this ignorance and misperception was a significant contributory factor leading up to the 1982 war. It remains significant today, and does not help in the reduction of tensions in the area.

This chapter looks solely at the geopolitics of the situation. For general reviews of the literature, see the very full survey by Etchepareborda (1984), and also Little (1984) and Tulchin (1987); for the history of the dispute, see Beck (1982) and Calvert (1982). After a brief outline of the nature and role of geopolitics in Argentina, this chapter looks at the Argentine and British views of the islands before 1982. It then examines the changes in geopolitical assessment that have occurred since the 1982 war, and discusses the current rivalries and tensions. The islands cannot be looked at in isolation, and the discussion examines how they are set in the regional context of the Falkland Islands Dependencies (South Georgia and the South Sandwich Islands), the South Atlantic, and the Antarctic sector. The chapter does not discuss the legal and political issues of sovereignty.

How to refer to the islands is itself a political problem. To insist on "Falklands" reads oddly when discussing the Argentine literature. Depending on context, the terms "Falklands," "Falklands/Malvinas," and (occasionally) "Malvinas" all appear in this chapter. None of this, however, should be taken as making political concessions!

ARGENTINE GEOPOLITICS

As in the neighboring South American countries of Brazil and Chile, geopolitical thought has a long history in Argentina. English-language surveys and reviews are provided by Child (1979, 1985) and by Pittman (1981), whose survey of the Argentine literature runs to over 400 pages. This literature expanded dramatically in the 1960s and 1970s as military governments came to power in the Southern Cone and the geopolitical literature in each of the countries reacted to that in the others. Within South America there has been a close association between geopolitics and the military, and in Argentina many of the major geopoliticians have been active, or (more usually) retired, military officers. The range of publications on geopolitics has probably been greater in Argentina than in any other South American country, and scores of books appeared in the period 1965 to 1982 (such as the works by Atencio, Briano, Cirigliano, Guglialmelli, Marini, Sanz, and Villegas), especially through the two publishing houses of El Cid and Pleamar in Buenos Aires. In addition, the work of the geopolitical institutes and their journals has been important. The Instituto Argentino de Estudios Estratégicos y de las Relaciones Internacionales (Argentine Institute of Strategic Studies and International Relations—INSAR), headed by General Guglialmelli, with its journal *Estrategia*, has been especially influential. From the mid-1970s, the Instituto de Estudios Geopolíticos (IDEG) and its journal *Geopolítica* augmented the scene, and in the early 1980s, General Goyret's glossily produced journal *Armas y Geoestrategia*. In addition, the international journal *Geosur*, journal of the Asociación Sudamericana de Estudios Geopolíticos e Internacionales (South American Association of Geopolitical and International Studies), published in Uruguay since 1979, carries many Argentinian articles.

The major theme of all this Argentine geopolitical literature is the search for a "national project," a spatial or territorial identity that will allow Argentina to recapture its former dynamism and leadership in South America. In Selcher's phrase, "Argentina remains a geopolitically dissatisfied country" (Selcher 1984). In the nineteenth century, Argentina lost (or failed to acquire) territory its government thought it should have inherited from the Spanish Empire. In the twentieth, it has been engaged in a wary territorial rivalry with Chile and, more particularly, with Brazil, with whom it has competed for political and economic leadership and for dominance over the La Plata Basin. Argentina's foreign policy has long had this strong territorial dimension, but the traditional approach was dominated by legalistic and historical perspectives. The appeal of geopolitics—both to the military and to a wider public—has been to set these goals in a wider strategic and power-political context as vital for the development of the nation. The impact of geopolitics has been widespread in Argentina, both through the books and journals, and through the many articles for the Buenos Aires press, such as *Clarín, La Nación, La Opinión, La Prensa*, and *La Razón*. Individual geopoliticians have had some role in the military governments, though not of the prominence of Pinochet in Chile or Golbery do Couto e Silva in Brazil. General Osiris Villegas has been interior minister, ambassador to Brazil, and negotiator in the Beagle Channel dispute, and Admiral Fraga has been social welfare minister and director of the Antarctic Bureau. The major role of geopolitics, however, has not been in directing specific policies in the corridors of power, but in molding the general climate of opinion, both of the military governments and the wider public. Since 1978, there has even been a geopolitical component in some school education (Pittman 1981).

This enormous geopolitical output has had a variety of geographical themes and emphases, and, indeed, one of the weaknesses of Argentine geopolitics has been its diversity and inability to settle on a single national thrust. But, as Child observes: "The fundamental characteristic of the Argentine school of geopolitics is its obsession with Brazil and its perception of Brazilian expansionism, primarily into the River Plate basin" (1979, 95). Other territorial and geographical concerns took second place to this characteristic. The date of Child's comment is important, though, for Argentine geopolitics was in a process of transition at this time, with a major shift of direction taking place toward the south and Antarctica in the late 1970s and early 1980s. It is this shift in attention that is central for the Falklands issue.

SOUTHERN GEOPOLITICS AND THE FALKLANDS/MALVINAS

The shift of geopolitical focus towards the south brought into prominence the role of the Falkland Islands (*Islas Malvinas*) and the other South Atlantic Islands of South Georgia, the South Sandwich Islands, South Orkneys, and Antarctica itself. These themes had not, of course, been totally absent from earlier Argentine geopolitical writing, and as early as 1916, Admiral Storni had drawn on Mahan and Ratzel to emphasize the importance of the sea in Argentine strategy (Storni 1916). But it is undoubtedly true that, up until the mid-1970s, they had not been a major focus. The continental, counter-Brazilian strategy was much more dominant.

One of the first signals of a shift was Cosentino's paper on the strategic value of the Falklands/Malvinas in *Estrategia* in 1970, followed by Hernández and Chitarroni's book on the Falklands/Malvinas (1977), and a whole series of other works on the South Atlantic and Antarctica (such as Fraga 1979, 1983; Guglialmelli 1983a; Milia 1978; Moneta 1983), together with many journal articles, especially in *Estrategia*.

No one factor explains this major geopolitical shift. It was the result of a series of interacting and reinforcing factors. Notable among them were: (1) the changing power balance with Brazil; (2) the disputes and negotiations with both Chile and Britain; (3) the military takeover in 1976; (4) the growing role of naval officers in geopolitical writing; and (5) the positive evaluation of resource potential in both offshore waters and Antarctica.

The first factor was that by the late 1970s the shift of relative power in favor of Brazil was becoming very obvious and, to Guglialmelli's disgust, the military government accepted this and moved towards a more cooperative posture, to gain advantages it could not win by competition and confrontation. The "southern project" was then an attempt to construct an alternative great future for Argentina in the Southern Cone.

The second factor was the negotiations with Britain during the 1970s over sovereignty of the Falklands/Malvinas, and the dispute with Chile over the Beagle Channel Islands, which came close to war in 1978. The political saliency of these issues contributed to geopolitical interest in the South, and, in turn, these issues were influenced by the geopolitical analysis: the effect was reinforcing, and lifted the stakes involved. *Estrategia* reported and commented on the Falklands negotiations in detail, bringing geopolitical analysis to bear, while the 1978 Beagle Channel dispute brought the Chilean geopolitician Pinochet into negotiation and confrontation with the Argentine geopolitician Villegas (Villegas 1978).

The third factor was the military takeover in 1976, which gave impetus and more influence to geopolitics, and put military and security objectives at the top of political priorities. In addition, the navy grew in geopolitical significance. Both in terms of political-military power and in terms of geopolitical writing, the Argentine navy had played a lesser role than the army, and this is one reason for the relative neglect of Storni's writings. By the 1970s, however, the Centro Naval and the Escuela de Guerra Naval (School of Naval War) were taking an active role in geopolitical seminars and publications (such as Cosentino 1970; Milia 1978; Fraga 1979, 1983). Also, institutional changes had allowed the navy to assume a more important role in the military system (Rouquié 1978; Coutau-Bégarie 1985).

Finally, there was the growing belief that the southern territories and offshore waters were not economically useless, but resource-rich. In part this reevaluation merely reflected the worldwide interest in marine resources (fisheries, offshore oil, and seabed resources, such as manganese nodules and other minerals), and the consequent interest in asserting and establishing maritime territories and boundaries (see Moneta 1983). But it also reflected particular regional circumstances, notably the possibility of substantial oil fields around the Falklands. Studies by the U.S. Geological Survey and a

British group suggested, on the basis of the sedimentary basin characteristics, that the oil deposits might be very large. The fisheries potential around the Falklands, and in particular the protein-rich krill found especially around South Georgia, was also being recognized. These speculations extended to Antarctica itself, and with the Antarctic Treaty System (which had suspended territorial claims for thirty years) up for possible review in 1991, Argentina was asserting its interest there also.

These changing views can be seen in Argentine geopolitical writing on the Falklands/Malvinas. Writing in 1970, Cosentino is strong on the strategic value of the location of the islands, but makes little mention of resource potential or territorial development. He traces the strategic history of the islands and their important role as a coal-bunkering port and key to the Cape Horn route before the opening of the Panama Canal in 1914. He then distinguishes two scales of strategic value: the islands' "local" significance to Argentina and the danger of hostile control over the approach routes to Argentine harbors, and then the wider significance of their South Atlantic location in the event of major international (or nuclear) war and the possible closure of the Panama Canal.

By the mid-1970s, resources, especially oil, were receiving emphasis, particularly in the aftermath of the 1973 oil price rise. Many assessments were wildly optimistic, seeing the *Mar Argentino* as a "new Kuwait." The British Shackleton report on the Falklands (1976) also discussed the oil potential, adding fuel to the flames. Thus Hernández and Chitarroni, in their *Malvinas: clave geopolítica* (1977), draw heavily on Cosentino for the strategic element, but also place much emphasis on oil resources, and the economic need for the recovery of the Malvinas. (See also Pocovi 1978, and Silenzi di Stagni 1982.)

Other studies set the Falklands/Malvinas issue in a much wider regional context: the whole chain of islands of the Southern Antilles, from the Falklands/Malvinas through to the South Orkneys, are the crucial link between Argentina and Antarctica. In his edited collection, Milia coins the term "*Atlantártida*" for the whole geopolitical space comprising the southwest Atlantic, the islands, and the Antarctic sector, seeing this as a coherent region, the "New Great Argentinian Frontier," which is both resource-rich and strategically important. In this and other studies (such as those of Fraga), control of the Falklands/Malvinas becomes a central element of a wider geopolitical strategy. These writings in the late 1970s notably upgrade the assessment of the region's strategic significance from that found in Cosentino. The rise of the supertanker, too large for the Suez Canal (and the closure of the canal after the 1967 Israel-Egypt war), had made the Cape of Good Hope route, and hence the South Atlantic, a key economic lifeline for Western Europe and the United States, which the expansion of the Soviet blue-water fleet put under constant threat. Adding to this the coming of the Alaskan supertankers, too large for the Panama Canal, and the possible threat to the Canal from left-wing control with the Panama Canal Treaty in 1977, Argentinian writers saw the South Atlantic as a key global strategic zone, and were constantly puzzled by the low priority given to the region by the United States and Western powers.

These geopolitical reappraisals significantly contributed to the political climate in Argentina in the years 1976 to 1982. They helped heighten interest in the recovery (or acquisition, from a British viewpoint) of the Falklands/Malvinas and associated territory: not only were the islands of great symbolic and historic significance for Argentina, but they were now also a central location in the territorial space Argentina needed for future economic and political development, the "New Frontier."

BRITISH PERSPECTIVES ON THE FALKLANDS

British perspectives on the Falklands were very different, and must be seen in the light of the long retreat from a worldwide imperial role after 1945. The post-1945 history of the islands and policy towards them is examined in Beck (1982) and Calvert (1982). The British withdrawal from "east of Suez" after 1960 included a gradual naval withdrawal from the South Atlantic (and an end to the South Atlantic Naval Command), culminating in the ending of the Simonstown naval agreement with South Africa in 1974. As the empire shrank to a scattering of islands and rocks, the Colonial Office in London became absorbed into the Foreign and Commonwealth Office (FCO), and eventually the Falkland Islands came under the Latin American section of the FCO. The Falklands were inevitably a remote, low priority; economic and political cooperation with Argentina was seen as important for future development. Postal, communications, and other service agreements were negotiated between the two governments (Beck 1982). Despite confrontations (such as that between the ships *Shackleton* and *Storni* in 1976), both Labour and Conservative governments in Britain continued to work at a negotiated settlement with Argentina, probably involving some sovereignty deal, throughout the 1970s.

To try to develop and diversify a declining Falklands economy based on wool production, the British government commissioned an economic study by Lord Shackleton to examine the economic potential of the islands. This discussed both offshore oil reserves and fisheries potential, and provoked furor in Buenos Aires, where it was seen as a reflection of British geopolitical ambitions in the area. In fact, the logic was otherwise: the interest in development was an attempt to help the islanders find a future, not a desire to retain the Falklands for Britain for their resource and geopolitical value. This is an important distinction.

Shackleton's report was, in fact, skeptical of both oil and large-scale fisheries development, partly because the oil reserves were highly speculative and very long-term, but also because they might alter the islands' character and way of life too dramatically. The economic and political commitment to implement Shackleton's recommendations was not entirely forthcoming; in particular, finance to build an extended runway (a Shackleton priority) was not found. Shackleton had recommended further cooperation with Argentina, and in the years that followed, Britain gave many signals of a willingness to negotiate and withdraw, if an acceptable political solution could be found. There is no doubt that both the government and the FCO looked to such an agreement, but no agreement acceptable to Argentina (which insisted on sovereignty transfer) could be sold to the islanders or the British House of Commons.

If Britain's appraisals of the resource potential of the Falklands were downbeat compared with those of the Argentine geopoliticians, the British strategic evaluation was almost nonexistent. As Coutau-Bégarie has commented in his study *Géostratégie de l'Atlantique Sud* (1985), the South Atlantic is "the forgotten quarter" in Anglo-American strategic writing, with virtually no discussion, in contrast to the North Atlantic, the Indian Ocean, and the Pacific. In fact, Coutau-Bégarie can point to only one English study of the Southern Oceans, that edited by the Conservative member of Parliament Patrick Wall (1977). This study does emphasize the Soviet threat to the oil routes, but the most telling point for the present argument is that the Falklands/Malvinas do not appear in either the index or the text. The emphasis is on the Cape of Good Hope route and the eastern South Atlantic; the Falklands are too far west to be relevant (see also Hurrell 1983). Equally, the worries about the Panama Canal have not materialized: the Canal remains open, and Alaskan oil is transshipped or piped across the isthmus (Coutau-Bégarie 1985). The revival of the Cape Horn route has not happened, and in both British and U.S. strategy the Falklands were very peripheral.

These are major differences between the British and Argentine perspectives. A further vital difference in viewpoint was the low, almost negligible, role of geopolitical thought and writing in Britain (and Europe and North America). Since 1945, geopolitics had been in disrepute, and it was not until the last decade that a revival has taken place (Hepple 1986). For British appraisal of the Falklands, this has meant not only that British writers (and policymakers) did not evaluate the islands' potential in geopolitical terms, but also that they remained largely ignorant of the vast Argentine geopolitical literature. Very little of the English-language (and especially British) literature refers to Argentine geopolitics. For example, the leading British political geographer, J. W. House, made no reference to any of this literature in his study of the conflict and its geography (1983). Even where writers are closer and more aware of the Argentine literature generally, they give it little mention, presumably seeing it in Anglocentric terms as of little significance. This is true of the leading British historian of the Falklands, P. J. Beck, in his studies of both the islands and Antarctica (1982,

1983, 1986). A notable exception is Little (1984), who comments briefly that "foreign policy making in Argentina is heavily conditioned by geopolitical considerations and these are not a preserve merely of the military" (p. 303). He also comments that neglect of this factor is one source of a common British puzzle: "why Argentina should want the Falklands at all is a bit of a mystery to some British observers" (p. 302). Makin (1983) has argued that the British government could have recognized from a whole series of signals in Argentina in early 1982 that an invasion of the Falklands was imminent. Equally, greater recognition of the key role given to the islands in the Argentine geopolitical literature would have led British commentators to take the growing crisis more seriously.

POST-1982: ARGENTINE PERSPECTIVES

The Argentine invasion of the Falklands on April 2, 1982, and the subsequent defeat by the British taskforce in June is a major watershed in geopolitical assessments. The Argentine defeat was a major setback for the "Southern Project," and the discrediting of the military, and the return to civilian rule under President Alfonsín, have had an impact on the role of geopolitics in Argentine political life.

All of the geopolitical journals devoted special or theme issues to the war, especially the causes of defeat and the British motives: *Estrategia 71-72* (April–September 1982), *Geopolítica 24, Armas y Geoestrategia* (2, (6) 1983), *Geosur* (35, 1983), and the Peruvian geopolitical journal *Estudios Geopolíticos y Estratégicos* (8, 1982). Geopolitics did not disappear after the 1982 war. In fact, a new journal began in 1982, as an offshoot of IDEG and *Geopolítica*: the *Revista Cruz del Sur*, official journal of the Instituto Latino-Americano de Estudios Geopolíticos (Latin American Institute of Geopolitical Studies—ILADEG), which attempts to develop a new role and perspective for geopolitics in post-1982 Argentina.

One of the most dominant themes of these reactions is the extreme surprise, not just at the British military response, but equally at the policy of the United States. This has had many effects in terms of reorienting Argentine geopolitics toward a strong emphasis on Latin American solidarity. There has, however, been no retraction from the earlier assessment of the importance of the Falklands/Malvinas in the South Atlantic. On the contrary, the 1982 war has reinforced that assessment. Again, the mismatch between Argentine perceptions and British perceptions leads to a major misunderstanding. The geopolitician's outlook can easily become such a mindset that the validity of other perceptions and motives is neglected or downgraded. So, if Britain fought for the Falklands (and the United States supported Britain), then it *must* be because Britain and the United States place a high geopolitical value on the islands and the region.

This interpretation of the war is widespread throughout the Argentine (and South American) geopolitical literature, including some of the most elegant essays. Villegas (1982) writes of the "apparent reasons" and "hidden interests" of the British action. These hidden interests are the regional resource potential, and the high strategic value (now revealed) that is placed on the area in terms of oceanic routes and access to Antarctica by Britain and its allies the United States and European Community. In a series of papers (1982a, 1982b, 1984) María del Carmen Llaver extends this view: the British-U.S. alignment is the result of NATO strategic interest in the South Atlantic, with the Falklands' value as a base. This has the potential to destabilize the South Atlantic, making it an arena of the cold war. In this, as in many other papers, a comparison is made with the Diego Garcia case: the British island in the Indian Ocean where 1,300 people were removed in order to lease it as a U.S. base, showing both the U.S.-British linkage and the hypocrisy of British claims of "self-determination." In a pair of witty, but also serious, papers, Boggio Marzet sets the analysis in a wider Atlantic context (1982, 1983), linking the South Atlantic islands to the islands of the Central Atlantic ridge (St. Helena, Ascension, Tristan da Cunha, and Gough). Two of these (Ascension and Tristan) already have a U.S. military presence. Militarization of the Falklands would extend the chain into the South Atlantic. With the possibility of these as island microstates dominated by Britain and the United States, this would be a process of "*caribización*," truly a "return of the Normans or North-men," the sea-pirates, to the South Atlantic. Such views are echoed by many writers, such as Quagliotti de Bellis (1982). At their most extreme, there are even suggestions that the whole 1982 war was a NATO trap, giving Britain and the United States an excuse to militarize the Falklands.

Although any "hidden conspiracy" is almost impossible to disprove, there is virtually no substance or evidence for these views. *Geosur* was able to translate and reprint an article originally published in a British defense journal by the British writer Tremayne (1982) that did see the strategic relevance of the Falklands to NATO, but this is a very atypical article, and certainly not representative of the policy motives in 1982. Britain fought the 1982 war for several reasons (including self-determination, resisting military aggression, and national pride), but a geopolitical scheme was not the motivation; nor was it the reason for U.S. support. If Britain had had geopolitical motives, it would not have failed to find the money to build a runway in the 1970s. (See also the interesting discussion of these issues by the French geographer/geopolitician Yves Lacoste, 1984.)

The response of Argentine geopoliticians to the 1982 defeat, and the new role of geopolitics in a civilian Argentina, needs some examination to help see the contemporary view of the Falklands/Malvinas. Geopolitical journals were ready to participate in debates on both the new international relations and the role of the military. Shortly before his death in June 1983, General

Guglialmelli edited a symposium on "La Crisis Argentina," published in *Estrategia* (73/74, Summer 1983), which included not only his own geopolitical assessment (Guglialmelli 1983b), but also a short paper by Alfonsín (1983). The new journal, *Revista Cruz del Sur*, began in the immediate aftermath of the war, with a degree of overstatement in its first editorial that captures the mood:

> ILADEG consider that the conflict in the South Atlantic constitutes a transcendent landmark in the life of the Southern Hemisphere. It is somewhat comparable to the fall of Constantinople, which marked the demise of one era and the birth of another. (vol. 1:1, August 1982.)

Substantive articles in the journal then examine the role of the military and its legitimate role in a civilian, democratic society (Ballester et al. 1983). Tibupena (1982) argues that the military, subordinate to the civil power and the constitution, have an important external role, but should not play an internal role in suppressing dissention. This is an important shift away from the linkage of geopolitics to national security doctrine and internal security, expressed most forcibly in Goyret (1980). The internal role of the Argentine military in the "dirty war" against all suspected dissention, and the reaction against the military after 1982, made it essential for the survival of geopolitics for geopoliticians to disown the internal security view. However, in the external role, the need to recover the Malvinas and other islands remains a major priority in Tibupena's scheme.

Ballester, García, Gazcón, and Rattenbach also reexamine the "Argentine national project" (Ballester et al. 1982). In their view, the policies adopted by the military government after 1976 were disastrous, with free market economics ruining the economy and dehumanizing society. As Tulchin (1984) comments "only under ideal economic conditions could the regime manage the contradiction between its economic and its geopolitical models." The post-1976 model could not deliver the economic growth needed for the military to attain its geopolitical goals. In their "basic doctrine," Ballester and colleagues reject both the Marxist approach and the model of liberal capitalism, arguing for an active state role in development and for close links with other Southern Cone countries. In all of this rethinking, however, the Malvinas and the South Atlantic remain a key geopolitical arena, and there is no withdrawal from earlier evaluations of this geopolitical space and its importance to Argentina.

Direct criticism of the whole philosophy of geopolitics appears to have been fairly limited (the qualification "appears" is an important one, since such critiques may have appeared in socialist or social science journals not available to this author in England). One exception is a paper by the population geographer Reboratti (1983). Drawing on recent work in human geography on the ideology of geopolitics, he criticizes the lack of any clear principles or methods in geopolitical analysis, its failure to relate to the social sciences, arguing that the obscurity of its methods and its oversimplification of society-space relations serves the interests of particular social groups (the military especially). Although the paper discusses the Argentine literature, its target is a very general one. A more particular critique is that found in Dabat and Lorenzano (1984). This is a provocative but stimulating Marxist analysis of Argentina and the role of the Malvinas. Interestingly, the authors' assessment of the resource and strategic role of the islands and the South Atlantic, and of British motivations, does not differ markedly from those expressed by a writer like Villegas. Where the difference really emerges is in Dabat and Lorenzano's questioning of the centrality of the "Malvinas strategy" for Argentina: "the recovery of the Malvinas by the Argentinian nation is a non-essential democratic claim" and "We must try more concretely to assess the significance of Argentina's territorial claim for the development of the nation" (p. 60). They continue:

> In the case of the Malvinas, we are faced with a just *territorial* claim, but one that is secondary to other social and political demands of the nation. The reintegration of the islands cannot solve any vital national needs in the long term, and much less so in the short term. (p. 61)

They conclude that the principal obstacle to development lies in the social relations of production, not lack of physical space: "Certainly the Malvinas are important. But of much greater importance is the recovery of the national territory for the people's use" (p. 62).

Dabat and Lorenzano's Marxist analysis is perhaps not very representative, but it finds an interesting echo in the very different work of Escudé (1986). Escudé argues that Argentina has paid a high economic price for its challenges to the U.S.–led international order, in contrast to Brazil's growth under pro-American and flexible foreign and economic policies. So he is arguing for a shift in Argentine policy in an opposite direction to most commentators (including the geopoliticians). He also argues that Argentina should reduce the relative role of territorial claims (including the Malvinas) in its foreign policies. These arguments (from very different political bases) that Argentina has overstressed the importance of territory and geopolitical space in its overall policy balance are likely to increase. The geopolitical literature of the late 1970s did exaggerate the significance of *Atlantártida* for the nation's development. Economic growth and prosperity is the prerequisite for any plans to exploit the resources of the harsh southern environment. The geopolitical analyses tended to reverse this causality, at the same time putting the military at the center of policy because territorial assertion, control, and security were at the core of their national project.

None of this means geopolitical perspectives will disappear, still less that Argentina will revise its claims to the Falklands/Malvinas. Geopolitics is more widely based than the military (who still retain an important role), and is increasingly integrated into a wider international-relations approach (as in Moneta's work). Space and territory are important, but not quite so important or central as Argentine geopolitical writings have suggested.

POST-1982: BRITISH REAPPRAISALS

The British victory in the Falklands in June 1982 has inevitably resulted in a reappraisal of the islands and their significance. Indeed, even before the taskforce had secured the islands, the British prime minister, Margaret Thatcher, had commissioned Lord Shackleton to update his report on economic development prospects (Shackleton 1982). The "Fortress Falklands" policy in the immediate aftermath of the war involved a major garrison, building the new airport and runway, and improving port facilities. The military costs have been enormous: for the period April 2, 1982 to 1984 (that is, including the actual war costs), it was over £2,000 million, or more than £1 million per inhabitant of the Falklands. These military costs will fall year by year, but even for 1987–1988 they are projected at £156,000 per inhabitant. In addition, economic development money has been found to implement Shackleton's recommendations, in contrast to the situation after the 1976 report, though there have still been criticisms that the new Falklands Development Corporation has been slow to act. Recommendations to diversify the economy and develop fishing have been accepted. The islanders' request for a 200-mile fishing zone was rejected, but a 150-mile zone has been set up (against Argentine protests), and many licenses granted. The 1982 Shackleton report still took a skeptical view of oil resource development: agnostic about the resources that might be there, and pessimistic about any useful development without renewed cooperation with Argentina.

An important aspect of the 1982 report, and of British policy since 1982, is its more regional approach, emphasizing the whole context of the Falkland Island Dependencies and British Antarctic Territory. In all earlier negotiations with Argentina, British thinking clearly separated the Falklands from the Dependencies (South Georgia and the South Sandwich Islands) and from British Antarctic Territory (south of 60 degrees, and so covered by the Antarctic Treaty). The legal and historical basis of the claims is very different (see Joyner 1984 for a detailed discussion). Argentina never drew the same distinction, but clearly saw the Falkland Islands as the key to the whole set, and, in fact, the first Argentine invasions took place in Southern Thule (South Sandwich Islands) in 1976, and in South Georgia. Since 1982, Britain has been much more aware of the Antarctic connection (Beck 1983, 1986). Funding for the British Antarctic Survey (BAS), which was under major threat in the late 1970s, has been dramatically increased. Equally, Shackleton (1982) comments: "South Georgia may in the long run be of greater importance to the future development of the potential wealth of the South-

West Atlantic and the Antarctic than the Falkland Islands" (p. 3). Here, Shackleton is thinking particularly of the tremendous krill concentrations around South Georgia. In all of this reappraisal there is a strong element of only really valuing something after it has nearly been taken away.

The result of the 1982 war is that Britain has entered into a high-cost policy of supporting both self-determination (to remain British) and economic development for the Falkland islanders. No other country (nor NATO) is going to share the base or its costs, and Britain will have to finance these alone. It is worth quoting Shackleton from a 1977 paper:

> Of course, it is perfectly clear that the Falkland Islands could develop much more rapidly through cooperation with Argentina, but it is also our firm view from a genuinely uncommitted stand-point that, with sufficient will and determination both on the part of the British government and the Falkland Islanders themselves, there is a real prospect of a good future for them.

For the foreseeable future, it is the latter option that is being followed. The question is whether Britain will sustain the necessary defense and economic commitment and expenditure into the years ahead. Although various groups and commentators discuss schemes of political compromise with Argentina (see Beck 1985), any settlement seems very unlikely in the near future. Neither side is willing to negotiate sovereignty. Foulkes (1983) has argued that the transfer of the islands to Argentina will occur peacefully when there is a combination of democracy in Argentina, a Labour government in Britain, and a Democrat in the White House. It would be very unwise to assume this. After the 1982 war, attitudes in Britain as well as in Argentina have altered. Although there is a wide diversity of viewpoint in Britain, it will be extremely difficult for any British government to sell a sovereignty deal to the islanders or to the British public. Unless economic and political cooperation can be discussed by Britain and Argentina without raising the sovereignty issue, then the impasse remains, and the southwest Atlantic will remain an area of tension.

CONCLUSIONS

This chapter has argued that the geopolitical perspective has been, and continues to be, an important but somewhat neglected aspect of the dispute over the Falklands/Malvinas. The British and Argentine writers and policymakers have had different perceptions, and have misperceived each other's motives and outlooks. One difficulty here is that it is nearly always possible to find some writer or representative espousing any particular viewpoint. Argentinian geopoliticians can find rightwing British Conservatives or U.S. Republicans to quote on the global strategic value of the Falklands, just as one can find left-wing Argentinian writers to quote on the demise of geopolitics. But neither extreme would be representative of influential opinion or of current policy, though, in the end, this assessment must be a matter of judgement. Overall, however, one can claim that one side (the British) has underrated, and indeed remained largely ignorant of, the geopolitical outlook of the other. Equally, Argentine geopoliticians have grossly overrated the geopolitical element in Britain's motives.

Since 1982, there have been changes taking place on both sides. The changes within Argentine geopolitics do not suggest any lessening of the key geopolitical role ascribed to the Falklands/Malvinas and the South Atlantic. But, under civilian rule the influence of geopolitics has diminished, and writers from both the left and liberal wings are beginning to argue that territorial claims and new geopolitical space should not be so central to Argentina's foreign policy or "national project." On the British side, the result of the 1982 war is increased attention and priority toward the Falklands Islands and the whole region. The new political and financial commitment means Britain must work toward a more coherent vision of the region, even if that is not expressed in explicit geopolitical language. The long-term future development of the southwest Atlantic clearly requires cooperation between Britain and Argentina, but with sovereignty the sticking-point for both sides, the rivalry looks set to continue. The South Atlantic will remain a zone of tension for some time to come.

chapter sixteen

Malvinas/Falklands, 1982-1988: The New Gibraltar in the South Atlantic?

Rubén de Hoyos

One of the consequences of the South Atlantic War of 1982 (be the war cause or pretext) was the emergence of a new geopolitical order in the South Atlantic. Like many other things that change, this new order also was similar to the past, although new priorities emerged. The United Kingdom retained its dominance and its century-and-a-half-long foreign insertion in the region; indeed, it has increased that presence. Argentina continued to gain diplomatic ground in international fora, where its claims are heard, but has been unable to exercise its local geopolitical presence with the success it desires. The United States has maintained its 150-year ambiguous juridical attitude toward the dilemma of the Malvinas/Falklands case, and thus remains the indirect beneficiary of the conflict. In 1831, the apparently unauthorized intervention of the guns of the U.S. Navy's frigate *Lexington* opened the door for British occupation, and in the 1982 war, the United States provided direct and public military aid to the British so that the situation would not change. The United States thus retained, behind an apparent neutrality, a geopolitical and geostrategic option by proxy of the United Kingdom.

This chapter will address these new developments by focusing on the most evident of them: the "Gibraltarization" of the disputed islands of the South Atlantic. The study is divided into three parts: The first, using the peculiar neologism "Gibraltarization," will establish a possible framework to explain how given choke points retain their significance and power by inserting, or reinserting, themselves into the prevailing system. The "Gibraltarized" choke point needed the system, and was needed by the system; as such, it has become the very symbol of the system itself. The second will examine the changes through which the islands have become "Fortress Falklands" by design of Prime Minister Margaret Thatcher's policy, and are thus the new Gibraltar of the South Atlantic during the years under consideration (1982–1988). The final part evaluates the possible geopolitical

Map 16.1   Malvinas/Falklands: An Overflow of NATO?

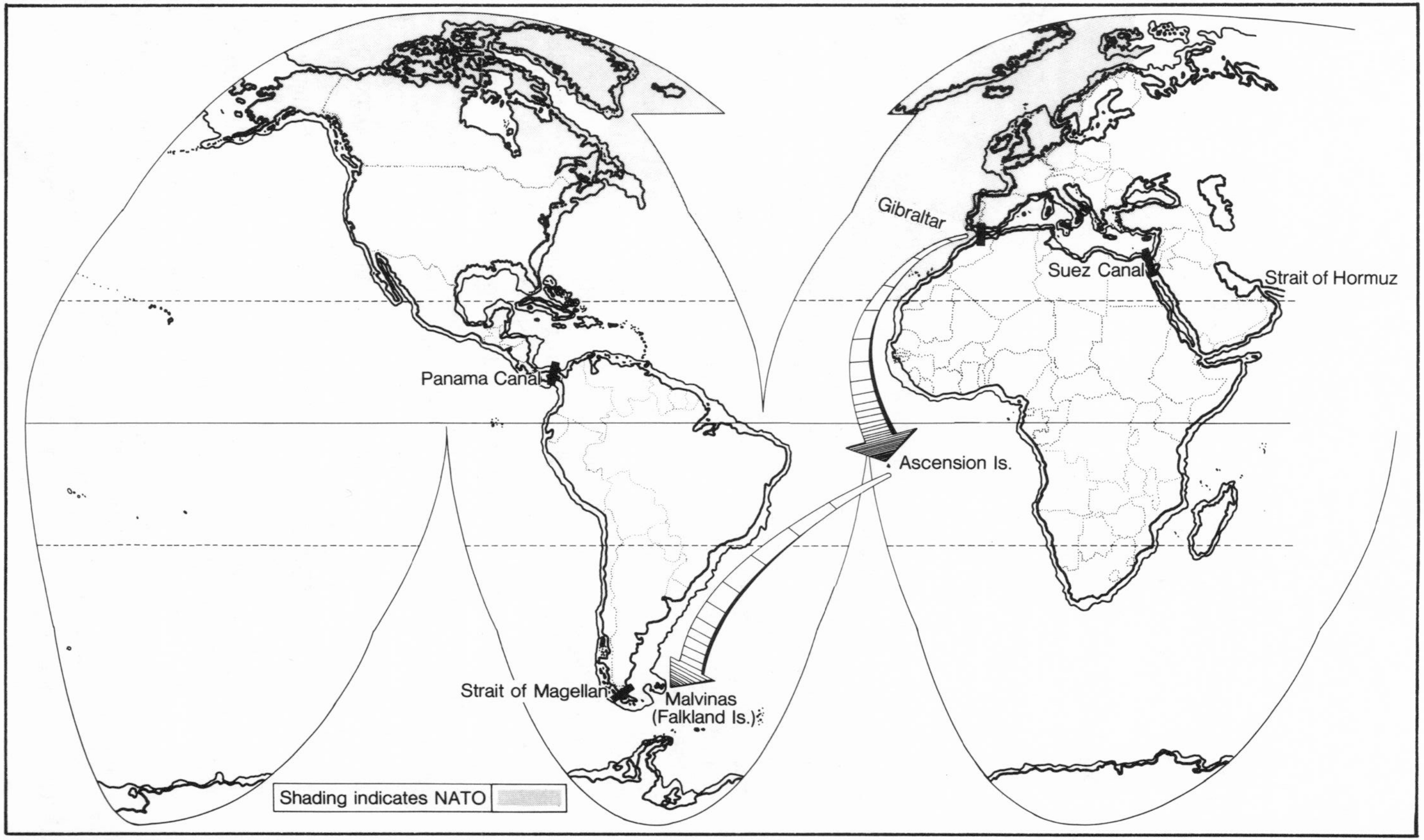

Gibraltar

Strait of Hormuz

Suez Canal

Ascension Is.

Panama Canal

Malvinas
(Falkland Is.)

Strait of Magellan

Shading indicates NATO

rearrangements which the Gibraltarization of the Malvinas/Falklands may have produced, and the political implications for the nations directly or indirectly involved.

GIBRALTAR AND GIBRALTARIZATION
AS A GEOPOLITICAL SYSTEM

"Gibraltarization" is the action by which something becomes "Gibraltar-like." Of all the characteristics involved, one stands out: the "Rock's" solid permanence and inexpugnability. That impregnability stems from its defenses, and hence the idea of "Fortress Falklands" links naturally to the action of Gibraltarizing the islands.

The Spaniards lost the Rock of Gibraltar to the Maghreb commander Tarik ibn Ziyad in 712, and, with the Rock, they also lost close to three-quarters of the Iberian peninsula for the next seven centuries (712–1462). When he took the Rock, Tarik made sure to erect a fort on its best lookout. In the year 1704, the Spaniards again lost the Rock due to inadequate defenses. Levie (1985, 8–12) has chronicled the events that provoked the British generals and admirals gathered in their war council aboard Admiral Sir George Rooke's flagship, and suggests that the principal reason for British success was the lack of Spanish defenses in Gibraltar. In the words of Oliver Cromwell, this allowed the British to create a permanent "annoyance to the Spaniards." In 1605, Cromwell was the first Englishman to see that control of the Mediterranean would require the possession and defense of the Rock of Gibraltar.

But perhaps a more immediate reason for Admiral Rooke's action was to save face. Rooke's armada had first attempted (and failed) to capture the city of Toulon (France), and then the city of Barcelona (the attack was repelled). He also attempted to capture the city of Cadiz, and was finally defeated. The eighty or so Spaniards defending the Rock were no match for the guns of the British Royal Navy and the 2000 seasoned marines who landed the morning of July 21, 1704.

And so began the British presence that has lasted nearly three centuries. The Moors placed a castle on the Rock but the British transformed the Rock into a castle, thus launching the myth of inexpugnability. They methodically improved its defenses, siege after siege, and war after war.

When the Mahgreb took Gibraltar (the closest point to their own shores), they poured onto the Iberian peninsula and created an imperial system whose key was the Rock. When the British took Gibraltar, it became the key to controlling the entrance to the Mediterranean. Later, the former Roman *Mare Nostrum* became a British channel toward their imperial possessions in the Middle East, India, and the Far East.

Let us retrace the process. Gibraltar became the key to two historical imperial systems: the Islamic Empire, which came to capture the Iberian peninsula, and the British Empire, spreading first into the Mediterranean and then later the world.

It is not that the Rock became the cause of the systems' growth, but an interesting geopolitical symbiosis is observable in this example. The fortress Rock became involved historically in the process of triggering both imperial systems, became their geostrategic key, and, finally, the symbol of the system. During World War II, Winston Churchill sent a special commando mission to North Africa with orders to return with more monkeys to restore the declining monkey population of Gibraltar. (There is a vague, but apparently widely believed. legend that with the death of the last monkey on the Rock the British presence will end).

The reading of contemporary British literature on Gibraltar emphasizes the perceived close link between the symbol, the empire that once covered a fifth of the earth's surface, and England itself: "while the Rock remains British, there will be an England."

Since the 1950s, the resourceful British have apparently found a way of keeping the Rock British, despite decades of steady imperial disintegration. This was done by transferring the geopolitical qualities and values of Gibraltar from the former British imperial system into the U.S.-European politico-military system of the North Atlantic Treaty Organization.

Brigadier C. N. Barely, writing in *Army Quarterly* (1967), concludes: "Even if Britain wished to surrender her rights she could not do so as a member of N.A.T.O.—bearing in mind the part which Gibraltar plays in N.A.T.O. defense." While the brigadier's reasoning sounds juridically flawed, it is still politically significant, particularly since the other party, the Spaniards, link Gibraltar and Spain's membership in NATO in a radically different way.

The Spanish administration of Felipe González has insisted that the reward for Spain's participation in the military system of NATO should be the return of Gibraltar to Spain, since it is self-defeating for Spain to consort militarily with the party that keeps a portion of Spain's territory by force.

THE PRESENT GIBRALTARIZATION OF THE MALVINAS/FALKLAND ISLANDS

If we understand "Gibraltarization" to be the process by which the islands are supposed to become the inexpugnable key to the security system and its symbol, then the Malvinas/Falkland islands, under a heavy process of militarization, are definitely being "Gibraltarized." This is not simply a matter of speculation, since it has been authoritatively stated by none other than Prime Minister Thatcher of the United Kingdom.

It is also a three-billion-pound fact, since this is the amount spent over the first three years the Thatcher government used to transform the Falkland Islands into the "Fortress Falkland Islands."

In January 1983, on the occasion of the Franks Committee's exculpatory report, Prime Minister Thatcher again stated that the solution to the Malvinas/Falkland Islands was the "Fortress Falklands" approach, since, as she affirmed, "it is not negotiable otherwise." "There is no other alternative" to the situation emerging from the Argentine landing on April 2, 1982 ("Falklands/Malvinas" 1983).

As reflected in the talks between the United States and Spain before, during, and after the 1984 visit of President Ronald Reagan to Madrid to negotiate the lease of U.S. military bases in Spain, a similar pattern has emerged. The Spanish administration seemed to hold the same misplaced hope held in 1982 by Argentine generals, and which was somehow still found in the expectations of the Alfonsín administration: that in the resolution of the Gibraltar or Malvinas dispute the role played by the United States is capital, since its influence on the occupant of 10 Downing Street could be decisive. In this, perhaps, they were right—or, to put it in other words, they may not be totally wrong.

The Thatcher government's policy for several years was denounced with vigor by her domestic opposition, which feared the economic and political costs of what former Prime Minister James Callaghan classified as an immediate "military victory," and as "a political defeat without exit" in the longer range ("Franks Attaches No Blame . . . ," 1983). This British policy, as announced to the international community by Thatcher, has been denounced equally by President Raul Alfonsín of Argentina at the General Assembly of the United Nations, by the members of the Organization of American States, by the Non-Aligned Nations in New Delhi (India), by European leaders, and by both the U.S. Congress and president.

The technical reasons for the decision argued by the British premier range from the United Nations principle of self-defense (Article 51) to the commitment of respecting "the wishes of the islanders." This is the stand the United Kingdom seems to have decided to pursue even in the face of United Nations Resolution 1514, which demands in Paragraph 5 that the parties consider only "the interests of the islanders" (the term "wishes" having been discussed out of the final text).

Neither the "wishes" nor the "interests" of the native inhabitants of the island of Diego Garcia were consulted when they were summarily transported to another island, so that the United Kingdom would be able to lease the island to the U.S. Navy and allow it to be transformed into a principal base.

Possibly neither the "interests," and certainly not the "wishes," of the inhabitants of Hong Kong, were consulted by Thatcher before signing the Anglo-Chinese treaty on the future of the crown colony after secret negotiations ("A Great Event . . . ," 1984).

This is noted not to impugn (or condone) Thatcher's policies but only to clarify the elasticity of apparently inflexible principles under the impact of "permanent interests," since a change along these lines could be one of the ways out of the South Atlantic impasse for a later British administration (but most certainly not during the lifetime of Margaret Thatcher).

Given that the Gibraltarization of the islands ("Fortress Falklands") is now the official policy, the second step is to record the fulfillment of the policy, to gauge the intensity with which it is being carried out, and thus to project its effects on the parties involved and on geopolitical regional balances.

We will attempt to present as briefly and objectively as possible a summary of the changes produced by this process of Gibraltarization. This will be accomplished by considering selected items and noting what the British policy has and has not been, both before and after the Argentine landing of April 2, 1982.

Budget Gibraltarization

As was noted above, a budget of three billion pounds was allocated for this purpose after 1982. And before that? It should be remembered that the islands were as remote from the budgetary priorities of home rule as they were geographically remote. A repeated complaint of the Kelpers (the citizens of the Falkland Islands) during the 1960s and 1970s was that what the United Kingdom regularly allocated to the islands was repaid by two million pounds' worth of hard currency derived from the Falkland Islands Company wool sales ("The Falkland Islanders" 1976). Of that allocation, only an insignificant percentage was devoted to "defence" (mostly police operations).

The same source (*The Times*) also carried the official announcement that, since the invasion, the United Kingdom has spent "one million pounds per inhabitant of the islands" ("Falklands Cost £ One Million" 1984). The civilian population of the islands has remained stable at around 1400 to 1500.

Military Gibraltarization

After the South Atlantic War, some 3,800 professional soldiers were stationed on the islands, and this number does not include Royal Air Force or Navy personnel.

Writing in *The Times of London* in early 1981, defense correspondent Henry Santhope, in commenting on the recently-created British Rapid Deployment Force (BRDF), concluded that, despite money and manpower constraints, such a BRDF should be trained for the Falkland Islands. The BRDF had been created in the United Kingdom following President Carter's original concept.

Before April 1982, there was a proudly expressed belief in high levels of the British Parliament that the thirty-seven Royal Marines stationed in Port

Stanley would be more than enough to deal with the whole Argentine military, if the need should arise.

Gibraltarization of Maritime Linkages

Before the events of April 1982, communications between the United Kingdom and the islands were handled by a small ship operated by the royally chartered Falkland Island Company. There were four trips a year, and each one took approximately a month.

The survey ship HMS *Endurance* was for years the only visible sign of official concern for the Falkland Islands. To the dismay of the Kelpers, the British decided to phase out the HMS *Endurance* in the 1980s, without apparent replacement. This led to a dangerous Argentine misperception that British rule was also being phased out.

After the South Atlantic War, however, things changed rapidly. On April 6, 1984, the authoritative *Latin American Weekly Report* (in an article titled "Argentina Show of Strength is Delayed" 1984) called its readers' attention to a "little reported fact": that NATO's second largest navy (that of the United Kingdom) had committed one-fourth of its surface ships to "that unstrategic South Atlantic bastion" (the Falkland Islands). This referred to surface ships only, and did not include the nuclear submarines that roamed the waters of the South Atlantic under a very creative reinterpretation of the Tlatelolco Treaty proscribing nuclear weapons in Latin America (whose Article 2 prohibits the use of nuclear ships with "warlike intention" in the area of the Rio Treaty).

Air Gibraltarization

Before April 1982, the only regular airport ever built on the Falklands was constructed and managed by the Argentine Air Force. The only air link for civil transport of persons, goods, and mail, and for emergencies was entirely an Argentine responsibility, in accordance with the July 1971 Anglo-Argentine agreement on air, sea, and postal communications.

The accord itself was another source of disappointment for the Kelpers (even if they benefited directly from it) because it made them more dependent on Argentina and eliminated their dream of having an international airport built by the United Kingdom.

After the South Atlantic War, it became a priority for the Thatcher administration to build a military airport able to receive the largest transport aircraft available. This airport was inaugurated ahead of schedule in April 1985. Previously, because of complicated in-flight refueling arrangements, each trip from London to the Malvinas/Falklands cost the British taxpayers approximately 450,000 pounds. The new airport is part of the air defense system that also includes squadrons of sophisticated Royal Air Force fighters and light bombers.

Before the events of 1982, the only radar was a conventional one typical of small airports. After the war, the British installed a radar that can reach out to the main Argentine naval base of Puerto Belgrano, over a thousand kilometers away, on the mainland.

The Exclusion Zone and Gibraltarization

The British blame the Argentines for the need to enforce the exclusion zone around the islands. The Argentines, for their part, seem to understand that the only condition the United Kingdom demands in exchange for cancelling the zone is for Argentina to give up, *de jure* and *de facto*, any right to the Malvinas. The Argentines also painfully learned not to believe Britannia when she waived her own rules by sinking the A.R.A. *Belgrano* in May 1982. But they understand that exclusion zones and Gibraltarization, despite the risks involved, are a way of bleeding the British administration in terms of both finances and public opinion.

We end this section with two observations: (1) Before April 1982, the islands were under the jurisdiction of the flamboyant civil governor Rex Hunt, and yet after the South Atlantic War, which was fought "for the wishes" of the Kelpers, Governor Hunt found himself under the jurisdiction of a miliary governor. (2) Before April 1982, the British Antarctic Survey was fading away, but after the South Atlantic War it suddenly started to recruit all types of researchers, suggesting that one of the possible consequences of the South Atlantic victory went beyond the political survival of the Kelpers and Mrs. Thatcher. There was also a new understanding and political will to project from the Malvinas/Falklands a new and vigorous British presence toward the Antarctic, which was about to start a thawing process, in political terms.

Having established that there was a Gibraltarization both by design and execution, this last observation serves as an introduction to an analysis of the possible consequences of Gibraltarization.

A GEOPOLITICAL REARRANGEMENT OF THE SOUTH ATLANTIC AS A CONSEQUENCE OF THE MALVINAS/FALKLANDS WAR?

"Fortress Falklands," be it wise or foolish, legal or not, is now a reality. There is a new Gibraltar in the South Atlantic. What are its geopolitical and political consequences? How is it being perceived and discussed by the interested parties? We will address below the viewpoints of four of the parties: Argentina, the United Kingdom, the United States, and the Soviet Union. In analyzing the ongoing process of Gibraltarization of the Malvinas/Falkland Islands, we must ask where this policy leads, and what costs will have to be paid by all those concerned. Further, if the policy changes previously existing geopolitical arrangements, what is the intensity of the change, and has the Gibraltarization of the islands succeeded in securing British control?

Impact of Gibraltarization on Domestic Politics of Argentina and the United Kingdom

In Argentina, military defeat accelerated (if it did not cause) the devolution of power from the Junta to a civilian, democratically elected government. It thus allowed the new president, Raul Alfonsín, to proceed with the "de-Gibraltarization" of the previous military establishment by reducing the military budget, putting the generals on trial, possibly reshaping the army into an all-volunteer force, and moving military garrisons from Buenos Aires to the interior.

This delicate process of de-Gibraltarization as a factor in Argentine politics, both domestic and foreign, won for Alfonsín and Argentina a rebirth of the state of law, and the reestablishment of a certain degree of international prestige.

The effects of British Gibraltarization of the islands on Argentina's domestic politics remains unclear. President Alfonsín acted according to the dictates of his own personal convictions and popular common sense in Argentina in his attempts to peacefully resolve the conflict with the United Kingdom by negotiation. This had been the course pursued for 134 years, and, for a period of sixteen years (from 1966 to 1982), it had been a frustrating diplomatic experience.

Although applauded in all diplomatic fora, the Argentine commitment to renewed negotiations has not been heard by the essential party: the United Kingdom. If this policy does not produce results, we can conceive of a situation in which Argentine public opinion and U.S. congressional opposition would respond to this nonproductive policy by forcing the president to get results, hardening his position, and requiring him to prove that the demilitarization of Argentina's policies has brought some results from the United Kingdom.

Military victory in the Malvinas/Falklands conflict brought Margaret Thatcher what was perhaps her finest hour. Her resolve and her triumph were popularly shared by the British as a sort of compensation for so many years of not being what they once were. But her process of Gibraltarization of the islands has awakened serious questioning in the press and the political opposition. The *Sunday Times* (Sciarli 1984) has pointed out that the cost of Gibraltarization represents nearly seven per cent of the military budget of the United Kingdom. Other estimates project total expenses of nearly six billion pounds in the six years after the war.

But, regardless of the failure of the Thatcher administration to provide jobs for the many unemployed in Britain, it appears that this level of military spending has not affected the overall popularity of Prime Minister Thatcher.

Impact of Gibraltarization on the Foreign Policies of Argentina and the United Kingdom

In Argentina, the inflexible British military policy of Gibraltarization has provided the Alfonsín government with a contrast to its own policies of negotiation, which have been welcomed internationally. This has been reinforced by the resolution of the old Argentine-Chilean dispute over the Beagle Channel, which thus reduced the Alfonsín administration's problems from two to one: Malvinas.

But Gibraltarization raises a serious challenge to Argentine claims of geopolitical hegemony in its immediate Atlantic Ocean (*Mar Argentino*). There are now three large military airstrips in the Falklands; the two largest, Mount Pleasant and Port Stanley, have runway lengths of 9,500 and 7,000 feet, respectively. The largest transport aircraft can safely use them even when fully loaded (Peters 1983). Suddenly, "Fortress Falklands" is perceived by Argentines and South Americans not simply as an alleged defense bastion, but as an offensive one as well.

The Gibraltarization of the islands and the enforcement of the exclusion zone by British planes and warships against the planes and ships of Argentina could one day lead to an accident that would affect both governments and re-ignite the conflict. Last but not least, Gibraltarization may sooner or later affect the long-standing Antarctic claims and projection of both nation's foreign policies toward that continent.

For the United Kingdom, the first consequence of the process of Gibraltarization is to guarantee by force British permanence on the islands. Great Britain seems to be moving away from archaic colonialist schemes by inserting the islands into the larger framework of defense of the free world via the North Atlantic Treaty Organization. In fact, the United Kingdom was able to achieve this goal during the South Atlantic War by diverting human and material resources allocated to NATO for the remote South Atlantic stage.

But perhaps more noteworthy at that time was the full endorsement of these actions by the United States, which took over British responsibilities in Europe; transferred U.S. oil and weapons to satisfy British needs in the Falklands/Malvinas War; and promised via personal assurances by Ronald Reagan that it "would not allow England to be defeated in the South Atlantic," pledging to replace losses ship for ship (*Latin American Weekly Report* 1984).

But what was initially simply a temporary *de facto* diversion of NATO resources to the South Atlantic has recently been submitted to a process of legitimation. No other meaning can be given to the statements of Margaret Thatcher in speaking to the members of the European Atlantic Community ("Margaret Thatcher Dinner Speech" 1984) when she declared the intention of her government was to "ignore the arbitrary boundary" of the North Atlantic (a possible reference to Article 6 of the NATO Treaty), and of her decision to operate "out of the NATO area" in defense of NATO's goals, for which she cited the Falklands/Malvinas case.

The process of Gibraltarization is now reaching the closure phase: from fortification to insertion into the most appropriate security system.

The Impact of Gibraltarization on the United States

What are the consequences of the fortification of the islands and their insertion into a security system in which the United States is a major partner? The behavior of the United States is a function of the tendency of contemporary superpowers to somehow become involved in all conflicts. The United States first volunteered, and was later requested, to take the role of mediator; the Argentine military demanded that Alexander Haig do it. And he, the former commander in chief of NATO, indeed did it.

During the conflict, the United States found itself caught between its obligations to NATO and the Rio Treaty. Its decisions and choices are now history. There were at one time other expectations on the part of the Argentines, based on the recollection that in 1956 a U.S. president had blocked the capture of the Egyptian Suez Canal by British forces. There was also a certain doubt in British circles precisely because of this same recollection.

The results were predictable. Historically speaking, the United States does not have a foreign policy per se toward Latin America or any of its nations. The behavior of the United States is a function of the given U.S. perception of its *global* aims and responsibilities, and not the result of bilateral or regional goals. For the Reagan administration, global geopolitical goals are identified with geostrategic ones, so much so that, at times, foreign policy seems like a process of world militarization. This could be a logical consequence of a perception that World War III is unavoidable, or perhaps because the administration attempts to prevent it through deterrent power.

Justified or not, there is a mutual lack of trust among the Latin American nations toward the United States that makes it difficult for them to coordinate their policies and actions. There is no similar lack of trust between the United Kingdom and the United States, joined by a "special relationship" and historical associations stemming from fighting the same wars and pursuing almost the same global interests.

From this viewpoint, it was easy to predict that the United States would prefer to have the Falklands and their strategic projection into the South Atlantic under the discreet influence of the British rather than that of the unpredictable Argentines. Regardless of other realities, Washington policymakers could identify more closely with familiar British "principles" than with untested Argentine ones.

The process of Gibraltarization of the Malvinas/Falklands by the British could not have been accomplished outside of these assumptions.

Gibraltarization and the Soviets

This analysis of the Gibraltarization process on Argentina (its immediate recipient) would be incomplete without a consideration of the effects of Gibraltarization on its ultimate addressee, the Soviet Union. The United Kingdom is now able to keep the Malvinas/Falklands by inserting them in the present global vision of NATO, but in so doing has produced the globalization of the Warsaw Pact and its insertion in the Americas.

During World War I, the British transformed the islands into an active military base from which to mount the defeat of the German fleet under the command of Graf von Spee on December 8, 1914. During World War II, the islands were also used for joint military purposes by the United Kingdom and the United States. The islands have a built-in strategic importance with a potential that can be activated in time of war. The Germans ignored this in 1914 and were defeated.

The Soviets are keenly aware of developments in the South Atlantic. At the United Nations, their official denunciations of British nuclear weapons in the area can be surmised to be not only an easy way to earn the gratitude of Argentina (of whom they are now the principal buyer of grain), but also an active way of protecting their own security interests.

The Gibraltarization of the islands, regardless of whether the original purpose was local defense or global offense, has created a geostrategic vortex that the Soviets ignore at their own risk.

Immediately after the South Atlantic War, the Academy of Sciences of the Soviet Union published a small but significant volume on the causes and consequences of the war (*La Crisis de las Malvinas* 1982). With much gusto, its ten contributors hammer the point home that, while the capitalist nations raised the threat of Communist aggression in the Americas, the physical aggression in this case came from them and not the Soviet Union. A second theme of the book argues the historical reality of an ongoing conspiracy between the United Kingdom and the United States to keep the islands in the hands of the British. A recurrent third theme seems to be the expression of a wish (and its verification) in the form of the collapse of the Inter-American Security System.

For years, the United States attempted to create a counterpart to NATO in the South Atlantic. For decades, U.S. strategists considered that the best possible option would be to have a neutralized South Atlantic inside the perimeter of the Rio and Tlatelolco treaties. It is academic how much that neutralization would be respected in time of war. But Gibraltarization is the equivalent of a NATO rejection of a possible South Atlantic Treaty Organization, and has attracted unabashed Warsaw Pact activity in a formerly neutralized area. No matter what justification could be alleged, or reasonable explanations given, that is now the bottom line.

In early 1988, a new development loomed over the South Atlantic: the British launched a combined Royal Navy–Air Force maneuver ("Fire Focus") centered on the Falkland/Malvinas Islands, and presumably involving the British Rapid Deployment Force. In response, Argentina took this unsettling event to the Organization of American States, and other Latin American nations expressed their concern directly to the British government.

Although British motivations for conducting this type of operation are unclear, no observer could discount the political repercussions and the many questions raised. For example, the area of the operation involved not only territorial waters claimed by Argentina, but also the Rio and Tlatelolco treaties. Further, it seems doubtful that such an operation could be conducted without support from the U.S. base on Ascension Island. Such U.S. involvement carries with it a price in terms of its relations with Latin America as well as Argentina. The reaction of the Soviet Union is another consideration, as well as the domestic effects of the British maneuver on President Alfonsín's moderate foreign policy. One domestic effect would be to unleash a wave of nationalism and a demand for a reaction that would give the military a higher profile; Margaret Thatcher might have overplayed her hand internationally as well as domestically. It is one thing to create new Gibraltars to secure an unfolding empire, and it is quite another thing to cling to outdated symbols when new realities are emerging.

Bibliography

Abecía Baldivieso, Valentín. *Las Relaciones Internacionales en la Historia de Bolivia.* 2 vols. La Paz: Editorial Los Amigos del Libro, 1979.

Academy of Sciences of the Soviet Union. *La Crisis de las Malvinas (Falklands): Orígenes y Consecuencias.* Colección Ciencias Sociales Contemporáneas. Moscow: The Academy of Sciences, 1982.

Alfonsín, R. "Modelar la Democracia Argentina con un Movimiento Democrático, Mayoritario, Popular y Reformista." *Estrategia* 73/74 (1983): 43–47.

Alvayay Fuentes, Eduardo. "La Carretera Austral." *Revista de Marina* 97 (1980): 144–150.

Anderson, Jack. "Argentina Move May be Warmup for Antarctica." *Washington Post* (April 12, 1982).

Andreasen, José Christian. "La Soberanía no se Negocia." *Participar* 1:3 (August–September 1978): 509.

Antarctic Treaty System, An Assessment. Proceedings from the 1985 Beardmore Conference on Antarctica. Washington, D.C.: National Academy Press, 1986.

Arévalo, Oscar. *Malvinas, Beagle, Atlántico Sur-Madryn. Jaque a la OTAN/OTAS.* Buenos Aires: Anteo, 1985.

"Argentina Show of Strength Is Delayed by Last Minute Rescue." *Latin American Weekly Report* 84 (April 6, 1984): 1.

Aron, Raymond. *Paz e Guerra entre las Nações.* Brasília: Editora Universitaria, 1976.

Arregui, J. *¿Qué Es el Ser Nacional?* Buenos Aires: Editorial Plus Ultra, 1973.

Asseff, Alberto Emilio. *Proyección Continental de la Argentina.* Buenos Aires: Editorial Pleamar, 1980.

Atencio, Jorge. *¿Qué Es Geopolítica?* Buenos Aires: Editorial Pleamar, 1975.

Auburn, F. M. *Antarctic Law and Politics.* Bloomington: Indiana University Press, 1982.

Avila, Federico. *Bolivia en el Concierto del Plata.* Mexico: Editorial Cultura, 1941.

Axelrod, R. *The Evolution of Cooperation.* New York: Basic Books, 1984.

Axelrod, R., and Keohane, R.. "Achieving Cooperation under Anarchy: Strategies and Institutions." In *Cooperation Under Anarchy*, ed. Kenneth A. Oye, 226–254. Princeton, New Jersey: Princeton University Press, 1986.

Azambuja, Pericles. *Antártica: Historia e Geopolítica*. Rio de Janeiro: Livros Brasileiros, 1982.

Azraai, Z. "The Antarctic Treaty System from the Perspective of a State not Party to the System." In *Antarctic Treaty System, An Assessment*, 305–314. Proceedings from the 1985 Beardmore Conference on Antarctica. Washington, D.C.: National Academy Press, 1986.

Backheuser, Everardo. *A Estructura Política do Brasil*. Rio de Janeiro: Mendoça, Machado and Cia., 1926.

Badia Malagrida, Carlos. *El Factor Geográfico en la Política Sud-americana*. Madrid, 1905.

Ballester, H. P.; García, J. L.; Gazcón, C. M.; and Rattenbach, A. B. "Poder Militar y Poder Civil." *Revista Cruz del Sur* 3 (1983): 9–21.

Ballester, H. P.; García, J. L.; Gazcón, C. M.; and Rattenbach, A. B. "Un Proyecto Nacional Argentino." *Revista Cruz del Sur* 1 (1982): 9–13.

Barely, Brigadier C. N. "Gibraltar 711–1967." *Army Quarterly* (1967): 340–346.

Beaufre, André. *An Introduction to Strategy*. New York: Praeger, 1966.

Beck, P. J. *The International Politics of Antarctica*. London: Croom Helm; New York: St. Martin's Press, 1986.

Beck, P. J. "Preparatory Meetings for the Antarctic Treaty 1958–9."*Polar Record* 22 (September 1985): 653–664.

Beck, P. J. "The Future of the Falkland Islands; a Solution Made in Hong Kong?" *International Affairs* 61 (1985): 643–660.

Beck, P. J. "Britain's Antarctic Dimension." *International Affairs* 59 (1983): 429–444.

Beck, P. J. "Cooperative Confrontation in the Falkland Islands Dispute. The Anglo-Argentine Search for a Way Forward, 1968–1981." *Journal of Interamerican Studies and World Affairs* 24 (1982): 37–58.

Beltramino, Juan Carlos. *Antártida Argentina: su Geográfia Física y Humana*. Buenos Aires: Centro Naval, 1980.

Bemis, Samuel Flagg. *La Política Internacional de los Estados Unidos-Interpretaciones*. New York: Biblioteca Inter-Americana 11, Lancaster Press, 1939.

Blanchard, W. O. "Foreign Trade Routes of Bolivia." *The Journal of Geography* 22 (December 1923): 341–345.

Boggio Marzet, P. "El Conflicto del Atlántico Sur y el Regreso de los Normandos." *Geosur* 43 (1983): 42–54.

Boggio Marzet, P. "El Nuevo Espacio Marítimo Argentino; la Dorsal Atlántica Central y las Islas de Tristán da Cunha." *Geosur* 35 (1982): 33–38.

Bond, Robert. "Brazil's Relations with Northern Tier Countries." In *Brazil in the International System: The Rise of a Middle Power*, ed. Wayne A. Selcher, 123–141. Boulder: Westview Press, 1981.

Boscovich, Nicolás. "Geopolítica y Geoestrategia en la Cuenca del Plata." *Geopolítica* (Buenos Aires) 9 (1983): 17–24.

Boscovich, Nicolás. "Un Proyecto Regional Argentino y la Natural Salida de

Bolivia al Mar." *Estrategia* 6 (September/October 1974): 28–43.

Bowman, Isaiah. "Trade Routes in the Economic Geography of Bolivia." *Bulletin of the American Geographical Society* 13 (January 1910): 22–37; (February 1910): 90–104; (March 1910): 180–192.

Bravo, A. A. "Objetivos Nacionales y los Conflictos en el Atlántico Sur." *Geopolítica* (Buenos Aires) 29 (1984): 5–6.

Brasil, Navy. *Primeira Expedição Antártica Brasileira*. Rio de Janeiro: Block, 1983.

Brazil, Empresa Brasileira de Notícias. *O Brasil na Antártica*. Brasília: Empresa Brasileira de Notícias, 1983.

Briano, J. *Geopolítica y Geoestrategia*. Buenos Aires: Editorial Pleamar, 1966.

British Information Services. "Margaret Thatcher Dinner Speech to the European Atlantic Group, Guild Hall, July 11, 1984." (August 1984).

Bruckner, P. "The Antarctic Treaty System from the Perspective of a Non-Consultative Party to the Antarctic Treaty." In *Antarctic Treaty System, An Assessment*. Proceedings from the 1985 Beardmore Conference on Antarctica, 313–336. Washington, D.C.: National Academy Press, 1986.

Bunster, Enrique. *Chilenos en California*. Santiago: Editorial del Pacífico, 1964.

Calvert, P. *The Falklands Crisis: The Rights and Wrongs*. London: Frances Pinter, 1982.

Caminha, João Carlos Gonçalves. *Delimeamentos da Estrategia*. 3 vols. Rio de Janeiro: Bibliex, 1983.

Campos, Martín. "El Mutún; Hipótesis para Comprender un Enigma Latino-Americano." *Estrategia* 4 (March/April 1973): 49–63.

Campos Pardo, Oscar A. "Antártida Región Fría de Política Encendida, Casi Candente . . ." *Antártida* (Argentina) 12 (May 1982): 5–7.

Campos S., Julio. "Consideraciones Sobre el Panorama Político-Estratégico Mundial." *Memorial del Ejército* 5 (September/October 1950): 3–5.

Cañas Montalva, Ramón. "Chile el Más Antártico de los Países del Orbe." *Seguridad Nacional* 14 (1979): 89–118.

Cañas Montalva, Ramón. "Los Hombres y el Territorio en el Trascendente Devenir Geopolítico de Chile." *Revista Geográfica de Chile: Terra Australis* 21 (1971): 51–77.

Cañas Montalva, Ramón. "Trascendencia Geopolítica de Canal Beagle." *Revista Geográfica de Chile: Terra Australis* 18 (1960): 6–20.

Cañas Montalva, Ramón. "La Antártica: Visionaria Apreciación del General O'Higgins." *Revista Geográfica de Chile: Terra Australis* 14 (1956–1957): 5–21.

Cañas Montalva, Ramón. "El Pacífico, Epicentro Geopolítico de un Nuevo Mundo en Estructuración." *Revista Geográfica de Chile: Terra Australis* 12 (1955): 11–17.

Cañas Montalva, Ramón. "Misión o Dimisión de Chile en el Pacífico–Sur Antártico." *Revista Geográfica de Chile: Terra Australis* 10 (1953): 9–12.

Cañas Montalva, Ramón. "El Valor Geopolítico de la Posición Antártica de Chile." *Revista Geográfica de Chile: Terra Australis* 9 (1953): 11–16.

Cañas Montalva, Ramón. "Agrupación o Confederación del Pacífico." *Revista Geográfica de Chile: Terra Australis* 2 (1949): 15–20.

Cañas Montalva, Ramón. "Reflexiones Geopolíticas sobre el Presente y el Futuro de América y de Chile." *Revista Geográfica de Chile: Terra Australis* 1 (1948): 27–40.

Canessa Robert, Julio. "Visión Geopolítica de la Regionalización Chilena." Address delivered in Montevideo, Uruguay, June 8, 1979.

Carlés, Fernando J. *Algunos Aspectos de la Geopolítica Boliviana.* Buenos Aires: Facultad de Derecho y Ciencias Sociales, Instituto de Derecho Internacional, Publicación No. 2, 1950.

Casellas, Capitánde Navío Alberto. *Antártica: Un Malabarismo Político.* Buenos Aires: Instituto de Publicaciones Navales, 1981.

Castex, Humbert. *Théories Estratégiques.* Paris, 1937.

Ceresole, Norberto. "La Situación Brasileña y la Integración del Cono Sur." Buenos Aires: ILCTRI, 1986.

Ceresole, Norberto. *Democracia, Potencial Militar y Autonomía Estratégica.* Buenos Aires: ILCTRI, 1984.

Ceresole, Norberto. *La Viabilidad Argentina.* Madrid: Editorial Altalena, 1983.

Child, Jack. *Antarctica and South American Geopolitics: Frozen Lebensraum.* New York: Praeger, 1988.

Child, Jack. *Geopolitics and Conflict in South America: Quarrels Among Neighbors.* New York: Praeger, 1985.

Child, Jack. "Geopolitical Thinking in Latin America." *Latin American Research Review* 14 (1979): 89–111.

Chile, INACh (Instituto Nacional Antártico de Chile). "Chile Consolida su Soberanía." *Boletín Antártico Chileno* 4 (January–June 1984): 59–62.

"The Chile Connection." *New Statesman* 109 (January 2, 1985): 8–10.

Church, George Earl. *The Route to Bolivia via the River Amazon; A Report to the Governments of Bolivia and Brazil.* London: Waterlow and Sons, 1877.

Cirigliano, Gustavo. *La Argentina Triangular.* Buenos Aires: Publicaciones Humanitas, 1975.

Cohen, Saul Bernard. *Geografía y Política en un Mundo Dividido.* Madrid: Ediciones Ejército, 1980.

Coloane, Francisco. *Antártica.* Santiago: Andrés Bello, 1985.

Cosentino, B. O. "El Valor Estratégico de las Islas Malvinas." *Estrategia* 6 (1970): 76–87.

Coutau-Bégarie, H. *Géostratégie de l'Atlantique Sud.* Paris: Presses Universitaires de France, 1985.

Crawford, Leslie. "Por un Club Antártico Ibero Americano." *Geosur* (Uruguay) 33 (1982): 34–43.

Crawford, Leslie. *Uruguay Atlanticense y los Derechos a la Antártica.* Montevideo: A. Monteverde y Cia., 1974.

Cubillos L., Hernán. "La Defensa Naval del Continente." *Revista de Marina* 4 (July/August 1950): 12–16.

Dabat, A., and Lorenzano, L. *Argentina: The Malvinas and the End of Military Rule.* London: Verso, 1984. Expanded and revised translation of *Conflicto Malvinense y Crisis Nacional.* Mexico City: Teoría y Político, 1982.

de Castro, Therezinha. "Dinâmica Geopolítica do Monroismo." *A Defesa Nacional* 724 (March/April 1986).

de Castro, Therezinha. "A Paz Morna da Guerra Fria." *A Defesa Nacional* 722

(November/December 1985).

de Castro, Therezinha. "Bacia do Prata: Pólo Geopolítico no Atlântico Sul." *A Defesa Nacional* 721 (September/October 1985).

de Castro, Therezinha. "Geopolítica do Confronto." *A Defesa Nacional* 716 (November/December 1984).

de Castro, Therezinha. "O Atlântico Sul no Contexto Regional." *A Defesa Nacional* 714 (July/August 1984).

de Castro, Therezinha. "O Ser e o Não Ser do TIAR." *A Defesa Nacional* 713 (May/June 1984).

de Castro, Therezinha. "O Cone Sul e a Conjuntura Internacional." *A Defesa Nacional* 712 (March/April 1984).

de Castro, Therezinha et al. *Los Países del Atlântico Sul—Geopolítica de la Cuenca del Plata.* Buenos Aires: Editorial Pleamar, 1983.

de Castro, Therezinha. "La Crisis de las Malvinas y sus Reflejos." *Geopolítica* 26 (1983): 29–34.

de Castro, Therezinha. "O Caribe em Ritmo de Guerra Fria." *A Defesa Nacional* 710 (November/December 1983).

de Castro, Therezinha. "Relações Brasil-Estados Unidos em face da Dicotomia Norte-Sul e Leste-Oeste." *A Defesa Nacional* 706 (March/April 1983).

de Castro, Therezinha. *Atlas-Texto de Geopolítica do Brasil.* Rio de Janeiro: Capemi Editora, 1982.

de Castro, Therezinha. "O Brasil e a Bacia do Prata." *A Defesa Nacional* 704 (November/December 1982).

de Castro, Therezinha. "Antártica: Suas Implicações." *A Defesa Nacional* 702 (July/August 1982).

de Castro, Therezinha. "Uruguay: Polígono Geopolítico do Cone Sul." *A Defesa Nacional* 700 (March/April 1982).

de Castro, Therezinha. "Argentina: Terminal de Rotas do Atlântico Sul." *A Defesa Nacional* 698 (November/December 1981).

de Castro, Therezinha. "América Central–Caribe: Area Vulnerável do Hemisfério Ocidental." *A Defesa Nacional* 694 (March/April 1981).

de Castro, Therezinha. *Rumo a Antártica.* Rio de Janeiro: Livraria Freitas, 1976.

de Castro, Therezinha. "Problemas Bolivianos." *Atlas de Relações Internacionais* 2. Supplement to *Revista Brasilera de Geográfica* 30 (January/March 1968): 2–9.

de Castro, Therezinha. "A Questão de Antártica." *Revista do Clube Militar* (June 1956): 189–194.

de Wit, M. *Minerals and Mining in Antarctica: Science and Technology, Economics and Politics.* Oxford: Clarendon Press, 1985.

Diaz, Alejandro S. *Ensayo Sobre la Historia Económica Argentina.* Buenos Aires: Amarrortu Editores, Talleres Gráficos Diot, 1976.

Diaz, Bessone, Ramón Genaro. "Que Hay Detrás de la Recuperación de las Islas Malvinas, Georgias y Sandwich del Sur." *Futurable* 14 (1982): 18–41.

Diehl, Jackson. "Argentine Leader Sworn In, Vows No `Great Changes.'" *Washington Post* (July 2, 1982): A21.

Diehl, Jackson. "Buenos Aires Junta Faces Deep Division." *Washington Post* (June 23, 1982): A1–A2.

Domínguez Ruiz, Ricardo. *Las Relaciones Argentina–Estados Unidos, 1976–1980.* Mexico City: Facultad de Economía, Universidad Nacional Autónoma de México, 1983.

Echevarría D., Gloria. "La Controversia entre Chile y Argentina sobre la Región del Beagle, Origen, Desarrollo y Desenlace." In *Cientocincuenta Años de la Política Exterior de Chile,* ed. Walter Sanchez and Teresa Pereire L. Santiago: Editorial Universitaria, 1977.

"The Economist: The U.S. Military Aid to U. Kingdom During the South Atlantic War." *Latin American Weekly Report* 84 (April 6, 1984): 7.

Edwards Vives, Alberto. *La Fronda Aristocrática.* 4th ed. Santiago: Editorial del Pacífico, 1952.

Encina, Francisco A. *La Cuestión de Límites entre Chile y La Argentina desde la Independencia hasta el Tratado de 1881.* Santiago: Editorial Nascimento, 1959.

Escudé, Carlos. *La Argentina vs. las Grandes Potencias (El Precio del Desafío).* Buenos Aires: Editorial de Belgrano, 1986.

Escudé, Carlos. *¿La Argentina: Paría Internacional?* Buenos Aires: Editorial de Belgrano, 1984.

Espinosa Moraga, Oscar. *El Precio de la Paz Chileno-Argentino 1810–1969.* 3 vols. Santiago: Editorial Nascimento, 1969.

Espinosa Moraga, Oscar. *Bolivia y el Mar 1810–1964.* Santiago: Editorial Nascimento, 1965.

Estrategia. *Arbitraje Sobre el Canal Beagle.* Documental Series 3. Buenos Aires: INSAR, 1978.

Etcheparaborda, R. "La Bibliografía Reciente Sobre la Cuestión Malvinas." *Revista Interamericana de Bibliografía* 34 (1984): 1–52, 227–288.

Eyzaguirre, Jaime. *Breve Historia de las Fronteras de Chile.* 9th ed. Santiago: Editorial Universitaria, 1978.

Eyzaguirre, Jaime. *Historia de Chile.* 2 vols. 3rd ed. Santiago: Editorial Zig-Zag, 1977.

Eyzaguirre, Jaime. *Chile y Bolivia: Esquema de un Processo Diplomático.* Santiago: Editorial Zig-Zag, 1963.

"Falklands Cost £ One Million on Islanders." *The Times of London* (November 13, 1984): 2a.

"Falklands/Malvinas, Thatcher Closes Down the Options." *Latin American Weekly Report* (February 4, 1983): 2.

F.I.D.E. "La Promoción Industrial Patagónica." *Coyuntura y Desarrollo* 62 (1983): 52–56.

Fifer, J. Valerie. *Bolivia: Land, Location and Politics Since 1825.* Cambridge: Cambridge University Press, 1972.

Foulkes, J. *Los Kelpers en las Malvinas y en la Patagonia.* Buenos Aires: Corregidor, 1983.

Fraga, Admiral Jorge A. *La Argentina y el Atlántico Sur.* Buenos Aires: Pleamar, 1983.

Fraga, Admiral Jorge A. *Introducción a la Geopolítica Antártica.* Buenos Aires: Dirección Nacional del Antártico, 1979.

"Franks Attaches No Blame to Government." *The Times of London* (January 19, 1983): 4.

Frenchman, Michael. "The Falkland Islanders May Be No More Than Pawns in a Game Britain Can Not Win." *The Times of London* (January 20, 1976): 12a.

Gaddes, John Lewis, *Strategies of Containment: A Critical Appraisal of Postwar American National Security Policy.* New York: Oxford University Press, 1982.

Gamba, Virginia. *The Falklands/Malvinas War: A Model for North-South Crisis Prevention.* Boston: Allen and Unwin, Inc., 1987.

García Lupo, Rogelio. *Diplomacia Secreta y Rendición Incondicional.* Buenos Aires: Legasa, 1983.

Glassner, Martin Ira. *Bibliography on Land-locked States.* 2nd ed. Dordrecht: Martinus Nijhoff, 1986.

Glassner, Martin Ira. *Access to the Sea for Developing Land-locked States.* The Hague: Martinus Nijhoff, 1970.

Glassner, Martin Ira. "Bolivia and an Access to the Sea." Master's Thesis, California State University at Fullerton, 1964.

Golbery, do Couto e Silva. *Geopolítica do Brasil.* Rio de Janeiro: Editorial José Olympio, 1967.

Gorman, Stephen S. "Security, Influence, and Nuclear Weapons: The Case of Argentina and Brazil." *Parameters* 9 (1979): 52–65.

Goyret, J. T. *Geopolítica y Subversión.* Buenos Aires: Ediciones Depalma, 1980.

Grabendorff, Wolf. "Interstate Conflict and Regional Potential for Conflict in Latin America." *Journal of Interamerican Studies and World Affairs* 24 (1982): 267–295.

Graham, Bradley. "S. American Democracy Put to Test." *Washington Post* (October 19, 1987): A15, A20.

"A Great Event of Historic Significance." *Beijing Review* 40 (October 1, 1984): 14–16.

Greever, Judy Groff. "José Ballivián and the Bolivian Oriente; A Study of Aspirations in the 1840's." Ph.D. diss. Radcliffe College, 1953.

Guglialmelli, Juan Enrique. *Geopolítica del Cono Sur.* 3rd ed. Buenos Aires: El Cid Editor, 1983.

Guglialmelli, Juan Enrique. "La Crisis Argentina. Una Perspectiva Geopolítica." *Estrategia* 73/74 (1983): 9–30.

Guglialmelli, Juan Enrique. "La Guerra de Malvinas. Falsos Supuestos Políticos Conducen a la Derrota." *Estrategia* 71/72 (1982): 19–90.

Guglialmelli, Juan Enrique. *El Conflicto del Beagle.* Buenos Aires: El Cid Editor, 1978.

Guglialmelli, Juan Enrique. *La Cuenca del Plata.* Buenos Aires: Tierra Nueva, 1974.

Guimaraes, L. F. "Macedo de Soares: The Perspective of a New Member." In *Antarctic Treaty System, An Assessment,* 337–344. Proceedings from the 1985 Beardmore Conference on Antarctica. Washington, D.C.: National Academy Press, 1986.

Gumucio, Mariano Baptista. "Geopolítica de Bolivia; Mediterraneidad y Destino." *Geopolítica* (Buenos Aires) 9 (1983): 5–18.

Gumucio, Mariano Baptista, and Saavedra Weise, Augustín. *Antología*

Geopolítica de Bolivia. La Paz and Cochabamba: Ediciones Los Amigos del Libro, 1978.

Hanson, Earl P. *South of the Spanish Main*. New York: Delacorte Press, 1967.

Henriques, Colonel Elber de Mello. *Uma Visão de Antártica*. Rio de Janeiro: Biblioteca do Exército, 1984.

Hepple, Leslie. "The Revival of Geopolitics." *Political Geography Quarterly*, Supplement to vol. 5 (October 1986): S21–S36.

Hernández, P. J., and Chitarroni, H. *Malvinas: Clave Geopolítica*. 2nd ed. Buenos Aires: Castaneda, 1982.

Hill, Clarence A., Jr. "Atlântico Sul." Lecture given at the Brazilian Naval War School, Rio de Janeiro, June 26, 1970.

Hilton, Stanley E. *Brazil and the Great Powers, 1930–1939: The Politics of Trade Rivalry*. Austin, Texas: University of Texas Press, 1975.

Hoffman, Fritz L., and Hoffman, Olga Mingo. *Sovereignty in Dispute: The Falklands/Malvinas, 1493–1982*. Boulder: Westview Press, 1982.

House, J. W. "The Political Geography of Contemporary Events: Unfinished Business in the South Atlantic." *Political Geography Quarterly* 2 (1983): 223–246.

Hurrell, A. "The Politics of South Atlantic Security: A Survey of Proposals for a South Atlantic Treaty Organization." *International Affairs* 59 (1983): 179–193.

Ibsen Gusmão, Admiral Câmara. "A Antártica: Interesses Científicos e Econômicos do Brasil." *Cadernos de Estudos Estratégicos* 2 (July 1982): 22–23.

Ihl C., Pablo. "Delimitación Natural entre el Oceano Antártico y El Atlántico en Resguardo de Nuestra Soberanía sobre la Antártica y Navarino." *Revista Geográfica de Chile: Terra Australis* 9 (1953): 45–51.

Ihl C., Pablo. "Línea Geopolítica de Chile." *Revista Geográfica de Chile: Terra Australis* 8 (1952): 26–43.

Ihl C., Pablo. "El Mar Chileno." *Revista Geográfica de Chile: Terra Australis* 10 (1951): 10–54.

Institut du Pacifique. *Le Pacifique "Nouveau Centre du Monde"*. Paris: Berger-Levrault, 1983.

Instituto Latinoamericano de Cooperación Tecnológica y Relaciones Internacionales (ILCTRI). *Fuerzas Armadas y Democracia*. Buenos Aires: ILCTRI, 1983.

Inter-American Defense College. *Trabajo de Investigación: La Antártida*. Washington, D.C.: IADC, 1987.

Inter-American Defense College. "Importancia Estratégica del Pacífico Sur, Atlántico Sur, y la Antártida." *Revista del CID* 12 (1985): 10–65.

Jervis, R. *Perceptions and Misperceptions in International Politics*. Princeton, N. J.: Princeton University Press, 1976.

Jordán Sandoval, Santiago. *Bolivia y el Equilibrio del Cono Sudamericano*. Cochabamba: Editorial Los Amigos del Libro, 1979.

Joyner, C. C. "Anglo-Argentine Rivalry after the Falklands/Malvinas War: Law, Geopolitics and the Antarctic Connection." *Lawyer of the Americas* 15 (1984): 467–502.

Kain, Ronald Stuart. "Bolivia's Claustrophobia." *Foreign Affairs* 16 (July

1938): 704–713.

Kelly, Philip. "The Geopolitics of Brazilian Expansion: A New Era of Pax Braziliana?" *Texas Journal of Political Studies* 9 (1987): 77–100.

Kelly, Philip. "Escalation of Regional Conflict: Testing the Shatterbelt Concept." *Political Geography Quarterly* 4 (April 1986): 161–180.

Kelly, Philip. "Buffer Systems in Middle America." In *Buffer States in World Politics*, ed. John Chay and Thomas E. Ross, 67–84. Boulder and London: Westview Press, 1986.

Kempff Bacigalupo, Roland. "Un Paliativo al Problema de la Mediterraneidad de Bolivia a través del Sistema Hidrográfico de la Cuenca del Plata." *Revista Argentina de Relaciones Internacionales* 14 (January/April 1978): 77–81.

Kimball, L. "Report on Antarctica." In *International Institute for Environment and Development 1987 Report*. Washington, D.C.: IIED, 1987.

Kirkpatrick, F. *South America and the War*. Cambridge, England: Cambridge University Press, 1918.

Kjéllen, Rudolf. *Antología Geopolítica*. Buenos Aires: Editorial Pleamar, 1975.

Knox, George A. "The Living Resources of the Southern Oceans." in *Antarctic Resources Policy: Scientific, Legal, and Political Issues*, ed. Francisco Orrego Vicuña. Cambridge: Cambridge University Press, 1983.

Kohlhepp, Gerd. "Colonización y Desarrollo Dependiente en el Oriente Paraguayo." *Revista Geográfica* 99 (1984): 5–33.

Lacoste, Y. "En Guis D'éditorial: La Mer: Quatre Grands Changements Géopolitiques." *Hérodote* 32 (1984): 3–41. (English language extract available in *International Geopolitical Analysis: A Selection from Hérodote*, ed. Girot, P., and Kofman, E., 46–58. London: Croom-Helm, 1986.

Lacoste, Y. *Les Pays Sous-Développés*. Paris: Presses Universitaires de France, 1959.

Lambert, J. *América Latina: Estructuras Sociales e Instituciones Políticas*. Barcelona: Editorial Ariel, 1973.

Lanzarini, Rear Admiral Mario. "La Argentina y la Situación Mundial." *Revista de la Escuela de Guerra Naval* 17 (1982).

Leal, General Jorge. "La Antártida Sudamericana y Latinoamericana." *Revista Militar* 711 (July–December 1983): 14–17.

Leal, Colonel Jorge. "El Petróleo y la Antártida." *Revista del Círculo Militar* (Argentina) 697 (November 1974): 8-12.

Leone, Manuel Oscar. "El Mutún: Un Factor de Integración." *Estrategia* 8 (March/April 1976): 25–35.

Leoni Houssay, Colonel Luis A. "Pinochet: el Führer Sudamericano." *Revista de Temas Militares* (Argentina) 11 (September 1984): 5–20.

Levie, Howard S. *The Status of Gibraltar*. Boulder: Westview Press, 1983.

Levingston, General Roberto M. "Antecendentes, Negociaciones y Consecuencias del Tratado de Paz y Amistad Argentino-Chileno." *Revista Argentina de Estudios Estratégicos* 2 (January–March 1985): 28–29.

Little, W. "The Falklands Affair: A Review of the Literature." *Political Studies* 32 (1984): 296–310.

Llaver, M. del Carmen. "Las Incidencias del Conflicto Malvinas en el Subsistema del Atlántico Sur." *Geopolítica* 28 (1984): 35–46. (Also published in *Geosur* 51 (1984): 35–51.)

Llaver, M. del Carmen. "Las Superpotencias y la Política de Poder en el Atlántica Sur." *Geosur* 33 (1982):3–17.

Llaver, M. del Carmen. "Atlántico Sur: Su Relevancia, Económica, Geopolítica y Geoestratégica." *Geopolítica* 24 (1982): 82–95.

Luard, Evan. "Who Owns the Antarctic?" *Foreign Affairs* 62 (1984): 1175–1193.

Mackinder, Halford J. "El Pívote Geográfico de la Historia." *Antología Geopolítica*. Buenos Aires: Editorial Pleamar, 1975.

Mahan, Alfred Thayer. *The Influence of Sea Power*. Boston: Little Brown, 1897.

Makin, G. "Argentine Approaches to the Falklands/Malvinas: Was the Resort to Violence Foreseeable?" *International Affairs* 59 (1983): 391–403.

Maquieira, C. "Antarctica Prior to the Antarctic Treaty: A Political and Legal Perspective." In *Antarctic Treaty System, An Assessment*, 49–54. Proceedings from the Beardmore Conference on Antarctica. Washington, D.C.: National Academy Press, 1986.

Marini, José Felipe. *Geopolítica de la Desintegración Rioplatense*. Buenos Aires: Escuela Superior de Guerra Aérea, 1986.

Marini, José Felipe. "Estrategia de la Democratización." In *La Razón* (June 17, 1986)

Marini, César José. *La Crisis en el Cono Sur*. Buenos Aires: SACI, 1984.

Mármora, Lelio. *Migración al Sur: Argentinos y Chilenos en Comodoro Rivadavia*. Buenos Aires: Ediciones Líbera, 1978.

Marull Bermúdez, Frederico. "Chile: Geopolítica del Pacífico Sur." *Geopolítica* (Uruguay) 5 (April 1978): 27–34.

Mastrorilli, Carlos. "La Cuestión del Atlántico Sur." *Revista Defensa* (Madrid) 1977.

Mastrorilli, Carlos. "Brasil y la Antártica: La Tesis de Therezinha de Castro." *Estrategia* 43–44 (November 1976–February 1977): 112.

McCann, Frank. "The Brazilian General Staff and Brazil's Military Situation, 1900–1945." *Journal of Interamerican Studies and World Affairs* 25 (1983): 299–324.

Meira Mattos, Carlos de. *Estrategias Militares Dominantes*. Rio de Janeiro: Bibliex, 1986.

Meira Mattos, Carlos de. "Bacia do Prata ou Cone Sul?" *A Defesa Nacional* 699 (January/February 1982).

Meira Mattos, Carlos de. *Geopolítica e Trópicos*. Rio de Janeiro: Biblioteca do Exército Editora, 1980.

Meira Mattos, Carlos de. "Atlántico Sul: Sua Importância Estratégica." *A Defesa Nacional* 688 (March/April 1980).

Meira Mattos, Carlos de. *Brasil—Geopolítica e Destino*. Rio de Janeiro: Livraria José Olympio Editora, 1979.

Meira Mattos, Carlos de. *Uma Geopolítica Pan-Amazônica*. Rio de Janeiro: Livraria José Olympio Editora, 1977.

Meira Mattos, Carlos de. *A Geopolítica e as Projeções do Poder*. Rio de Janeiro: Livraria José Olympio Editora, 1977.

Meira Mattos, Carlos de. "O Poder Militar e a Política Internacional." *Revista*

Brasileira de Política Internacional 63–64 (1973): 63–80.

Meneses C., Emilio. "Estructura Geopolítica de Chile." *Revista de Ciencia Política* 1–2 (1981): 105–161.

Meneses C., Emilio, and Goddard Dufeau, German. "Fallo Arbitral del Beagle de Mayo de 1977: Informe Preliminar sobre sus Proyecciones Geopolíticas y Estratégicas." Internal Working Document, Catholic University of Santiago, Chile, 1978.

Menezes, Eurípides Cardoso de. *A Antártica e os Desafíos do Futuro.* Rio de Janeiro: Capemi Editora, 1982.

Menezes, Eurípides Cardoso de. "Os Direitos do Brasil na Antártica." Address, Joint Session, Congresso Nacional do Brasil, November 29, 1970.

Mercado Jarrín, General Edgardo, ed. *El Perú y la Antártida.* Lima: Instituto Peruano de Estudios Geopolíticos y Estratégicos, 1984.

Mericq, General Luis S. *Antarctica: Chile's Claim.* Washington, D.C.: National Defense University, 1987.

Milia, Admiral Fernando A., ed. *La Atlantártida: Un Espacio Geopolítico.* Buenos Aires: Pleamar, 1978.

Mitchell, B. *Frozen Stakes: The Future of Antarctic Minerals.* London: International Institute for Environment and Development, 1983.

Mitrany, D. *A Working Peace System.* Chicago: Quadrangle Books, 1943.

Mocelin, Jane S. P. *Antártida, o Sexto Continente.* São Paulo: Olivetti, 1982.

Moneta, Carlos Juan, ed. *Geopolítica y Política de Poder en el Atlántico Sur.* Buenos Aires: Editorial Pleamar, 1983.

Moneta, Carlos Juan. "Antarctica, Latin America, and the International System in the 1980's." *Journal of Interamerican Studies and World Affairs* 23 (1981): 29–68.

Montes, Oscar Antonio. *Message of the Argentine Foreign Minister Declaring the Nullity of the Beagle Arbitral Award.* January 25, 1978.

Morales, Waltraud A. *La Geopolítica de la Política Exterior de Bolivia.* Santiago: Academia de Humanismo Cristiano, Documentos de Trabajo PROSPEL 2, September 1984.

Morales, Waltraud A. "Bolivian Foreign Policy: The Struggle for Sovereignty." In *The Dynamics of Latin American Foreign Policies: Challenges for the 1980s,* ed. Jennie K. Lincoln and Elizabeth G. Ferris, 171–191. Boulder: Westview Press, 1984

Moreira, Luiz Carlos Lopes. *A Antártida Brasileira: Sonho ou Realidade?* Rio de Janeiro: Feplam, 1982.

Moro, Rubén O. *Historia del Conflicto del Atlántico Sur: La Guerra Inaudita.* Buenos Aires: Escuela Superior de Guerra Aérea, 1985 (also—a more accessible publication—Buenos Aires: Editorial Pleamar, 1985).

Musso, Julio. *Antártida Uruguaya.* Montevideo: El País, 1970.

O'Connor d'Arlach, Octavio. "Una Ruta Fluvial para el Sud de Bolivia y la Navegación del Río Bermejo." *Revista de la Sociedad Geográfico y de Historia "Tarija"* (Tarija, Bolivia) 5 (March 15, 1946): 5–9.

O'Donnel, G., and Link, D. *Dependencia y Autonomía.* Buenos Aires: Amorrortu Editores, 1973.

Orrego Vicuña, Eugenio. *Política Antártica de Chile.* Santiago: Universidad de Chile, 1984.

Orrego Vicuña, Eugenio. *La Antártica y sus Recursos.* Santiago: Editorial Universitaria, 1983.

Orrego Vicuña, Francisco, ed. *Política Oceánica.* Santiago: Editorial Universitaria, 1978.

Orrego Vicuña, Francisco, and Salinas Aray, Augusto, eds. *El Desarrollo de la Antártica.* Santiago: Editorial Universitaria, 1977.

Oye, Kenneth A., ed. *Cooperation under Anarchy.* Princeton: Princeton University Press, 1986.

Palermo, Vicente. *Espacio Americano y Espacio Antártico.* Buenos Aires: Instituto Antártico Argentino, 1979.

Parfit, Michael. *South Light.* New York: Macmillan, 1985.

Peters, John. "Report to the Defense Committee of the British Parliament." *The Times of London* (June 15, 1983): 4b.

Pinochet de la Barra, Oscar. "Evolución Política-Jurídica del Problema Antártico." *Geosur* 51 (1984): 11–22.

Pinochet de la Barra, Oscar. *La Antártica Chilena.* Santiago: Editorial Andrés Bello, 1976.

Pinochet Ugarte, Augusto. Interview with Colonel Howard T. Pittman. Santiago, Chile, at the Moneda, May 8, 1986.

Pinochet Ugarte, Augusto. *Geopolítica.* 3rd ed. Santiago: Editorial Andrés Bello, 1977.

Pinto Coelho, Arístides. *Nos Confins dos Tres Mares . . . a Antártida.* Rio de Janeiro: Biblioteca do Exército, 1983.

Pittman, Howard T. "The Impact of Democratization on Geopolitics and Conflict in the Southern Cone." Paper presented at the annual meeting of the Latin American Studies Association, Boston, 1986.

Pittman, Howard T. "Chilean Foreign Policy: The Pragmatic Pursuit of Geopolitical Goals." In *The Dynamics of Latin American Foreign Policies: Challenges for the 1980s,* ed. Jennie K. Lincoln and Elizabeth G. Ferris, 125–135. Boulder: Westview Press, 1984.

Pittman, Howard T. "Geopolitics in the ABC Countries: A Comparison." Paper presented for the Western Political Science Association, Seattle, Washington, 1983.

Pittman, Howard T. *Geopolitics of the ABC Countries: A Comparison.* Ph.D. diss. American University, Washington, D.C., 1981.

Pittman, Howard T. "Geopolitics in the ABC Countries: A Comparison." Paper presented at the Conference of Latin Americanist Geographers, Buffalo, New York, October 16–18, 1981.

Pocovi, A. S. "Hidrocarburos Bajo el Mar Argentino." *Estrategia* 49–50 (1978): 49–59.

Posen, Barry R., and Van Evera, Stephen W. "Reagan Administration Defense Policy: Departure from Containment." in *Eagle Resurgent: The Reagan Era in American Foreign Policy,* ed. Kenneth A. Oye, Robert J. Lieber, and Donald Rothchild, 75–114. Boston and Toronto: Little, Brown and Company, 1987.

Pozzo Medina, Julio. *Geopolítica y Geoestrategia.* La Paz: Editorial Don Bosco, 1984.

Qaasim, S. A., and Rajan, H. P. "The Antarctic Treaty System from the

Perspective of a New Member." *Antarctic Treaty System, An Assessment*, 345–374. Proceedings from the 1985 Beardmore Conference on Antarctica. Washington, D.C.: National Academy Press, 1986.

Quagliotti de Bellis, Bernardo. "Dinámicas en el Cono Sur." *Geosur* 63/64 (1986): 3–22.

Quagliotti de Bellis, Bernardo. "Inglaterra–Estados Unidos y las Malvinas: Un Nuevo Estado Tapón?" *Geosur* 34 (1982): 3–21.

Ramacciotti de Cubas, Beatriz. *El Perú y la Cuestión Antártica*. Lima: CEPEI, 1986.

Ratzel, F. *Politische Geographie*. Munich: Ed. Oldenburg, 1923.

Reboratti, C. E. "El Encanto de la Oscuridad: Notas Acerca de la Geopolítica en la Argentina." *Desarrollo Económico* 23 (1983): 137–144.

Reimann, Elizabeth. *Las Malvinas: Traición Made in USA*. Mexico: Ediciones Caballito, 1983.

Republic of Chile. *Memoria Anual 1984–85 y Plan de Actividades 1985–86*. Santiago: Instituto Geopolítico de Chile, 1986.

Republic of Chile. *Reseña*. Santiago: Instituto Geopolítico de Chile, 1986.

Republic of Chile. *Objetivo Nacional y Políticas Generales del Gobierno de Chile*. Santiago: Office of the President, 1981.

Republic of Chile. *Programa Socio-Económico 1981–1989*. Santiago: Ministerio del Interior, 1981.

Republic of Chile. *Declaración de Principios del Gobierno de Chile. Santiago, March 11*. Santiago: Junta del Gobierno, 1974.

Riesco, Ricardo. "Relaciones Geopolíticas en Sudamérica." *El Mercurio* (January 27, 1986): A2.

Riesco, Ricardo. "La Antártica, Ventana Abierta al Espacio Cósmico." *El Mercurio* (December 23, 1984): A2.

Riesco, Ricardo. "Geopolítica Austral y Antárlica." *Boletín Antártico Chileno* 4 (July–December 1984): 14–17.

Riesco, Ricardo. "Chile y sus Perspectivas Geográficas frente al Pacífico y la Antártica." *Revista de Geografía. Norte Grande* 7 (1980): 52–54.

Riviere d'Arc, Hélène. "Economie Frontaliere et 'Poles de Développement': El Mutún et Itaipú." *Cahiers des Ameriques Latines* 18 (1978): 25–32.

Rodriguez, Bernardo N. *Soberanía Argentina en la Antártida*. Buenos Aires: Centro de Estudios Estratégicos, 1974.

Rogers, William O. "The U.S. and Latin American Relations." *Foreign Affairs* 63 (1985): 560–580.

Rosch, Captain Peter. "Navy of the Federal Republic of Germany." *Europäische Wehrkunde* 3 (1984).

Rouquié, A. *Pouvoir militaire et société Politique en république Argentine*. Paris: Presses de la Fondation nationale des sciences politiques, 1978.

Rozitchner, Leon. *Las Malvinas: de la Guerra 'Sucia' a la Guerra 'Limpia'*. Buenos Aires: CEAL, 1985.

Russell, Roberto. "Relaciones Internacionales de Argentina, Brasil y México." *América Latina Internacional* 1 (1984): 29–32.

Rybakov, Y. "Jurisdictional Nature of the 1959 Treaty System." In *Antarctic Treaty System, An Assessment*, 33–45. Proceedings from the 1985 Beardmore Conference on Antarctica. Washington, D.C.: National Academy

Press, 1986.

Salgado, Alba, Admiral Jesús. "Política y Estrategia en el Atlántico Sur." Madrid: Instituto de Cuestiones Internacionales, 1984.

Salinas F., Ramón. "La Posición Estratégica de Chile en la Defensa del Continente." *Memorial del Ejército* 3 (October 1947): 12–15.

Samhaber, Ernst. *Sudamérica. Biografía de un Continente.* Buenos Aires: Editorial Sudamericana, 1945.

San Martín, Humberto. "La Siderurgia Argentina." *Carta Política* (Buenos Aires) (1976).

Santhope, Henry. "Progress Slow in Extending Capability of Armed Forces." *The Times of London* (March 3, 1981): 6.

Santis, Hernán. Letter to Howard T. Pittman. December 11, 1984.

Schneider, Ronald M. *Brazil: Foreign Policy of a Future World Power* Boulder: Westview Press, 1976.

Schultz, George. "La Argentina y el Conflicto Este-Oeste." *La Nación* (Buenos Aires), November 14, 1986.

Sciarli, John. "Fortress Falklands." *The Sunday Times of London* (June 15, 1983): 4b.

Selcher, Wayne A. "Relaciones entre Brasil y Argentina en la Década del 80: De una Cautelosa Rivalidad a una Competencia Amistosa." *Estudios Internacionales* 70 (April/June 1985).

Selcher, Wayne A. "Brazilian-Argentine Relations in the 1980s: From Wary Rivalry to Friendly Competition." *Journal of Interamerican Studies and World Affairs* 27 (1985): 25–53.

Selcher, Wayne A. "Recent Strategic Developments in South America's Southern Cone." In *Latin American Nations in World Politics*, ed. H. Muñoz and J. S. Tulchin, 101–118. Boulder: Westview Press, 1984.

Shackleton, Edward A.A.. *Falkland Islands. Economic Study 1982.* London: HMSO, 1982.

Shackleton, Edward A.A. "Prospect of the Falkland Islands." *Geographical Journal* 143 (1977): 1–13.

Shackleton, Edward A.A. *Economic Survey of the Falkland Islands.* 2 vols. London: Economist Intelligence Unit, 1976.

Shapley, D. *The Seventh Continent, Antarctica in a Resource Age.* Washington, D.C.: Resources for the Future, Inc., 1985.

Silenzi de Stagni, J. *Las Malvinas y el Petróleo.* Buenos Aires: El Cid, 1982.

Spate, Oscar H. K. *The Spanish Lake.* Minneapolis: University of Minnesota Press, 1979.

Spykman, Nicholas J. *America's Strategy in World Politics: The United States and the Balance of Power.* New York: Harcourt Brace and Company, 1942.

Sternberg, Hilgard O'Reilly. "'Manifest Destiny' and the Brazilian Amazon: A Backdrop to Contemporary Security and Development Issues." *Yearbook (1987) of the Conference of Latin Americanist Geographers* 13 (1987): 25–35.

Storni, S. R. *Intereses Argentinos en el Mar.* 1916. Reprint. Buenos Aires: Centro Naval, Instituto de Publicaciones Navales, 1967.

Subercaseaux, Benjamin. *Tierra de Oceáno.* Santiago: Ediciones Ercilla, 1951.

Subercaseaux, Benjamin. *Chile o Una Loca Geográfia.* Santiago: Editorial

Ercilla, 1940.

Tambs, Lewis A. "Rubber, Rebels, and Rio Branco: The Contest for the Acre." *Hispanic American Historical Review* 46 (1965): 254–273.

Tambs, Lewis A. "Geopolitical Factors in Latin America." In *Latin America: Politics, Economics and Hemisphere Security*, ed. Norman A. Bailey, 31–49. New York: Praeger, 1965.

Tanzi, Héctor J. "El Continente Antártico y la Tierra del Fuego en el Siglo XVI." *Revista de Historia de América* 100 (1985): 13–53.

Tibupena, H. "Aspectos del Nuevo Proyecto Nacional y Continental que Orientan la Confección de un Plan de Desarrollo y Seguridad de Emergencia." *Revista Cruz del Sur* 1, part 2 (1982): 9–19.

Travassos, Mario. *Projeção Continental do Brasil*. São Paulo: Companhia Editora Nacional, 1947.

Tremayne, P. "Las Islas Falkland." *Geosur* 35 (1982): 21–32.

Trias, Vivian. *Imperialismo y Geopolítica en América Latina*. Buenos Aires: Editorial Cimarrón, Libertad, 1973.

Troncoso D., Arturo. "Geopolítica del Pacífico." *Revista de Marina* 6 (November/December 1961): 24–37.

Tulchin, Joseph S. "Authoritarian Regimes and Foreign Policy: The Case of Argentina." In *Latin American Nations in World Politics*, ed. H. Muñoz and J. S. Tulchin, 186–199. Boulder: Westview Press, 1987

Tulchin, Joseph S. "The Malvinas War of 1982: An Inevitable Conflict that Never Should Have Occurred." *Latin American Research Review* 22 (1987): 123–141.

Tulchin, Joseph S. "Regímenes Autoritarios y Política Exterior: El Caso de Argentina." In *Entre la Autonomía y la Subordinación: Política Exterior de los Países Latinoamericanos*, ed. Joseph S. Tulchin and Heraldo Muñoz. Buenos Aires: Grupo Editor Latinoamericano, 1984.

Ugarte, M. *La Reconstrucción de Hispanoamérica*. Buenos Aires: Editorial Patria Grande, 1962.

United Nations, General Assembly, 39th Session. *Question of Antarctica*. 3 vols. New York: United Nations, 1984.

"U.S. Military Aid to Britain, The Cat Is Out of the Bag." *Latin American Weekly Report* 84 (March 16, 1984): 7.

Valencia Vega, Alipio. *Geopolítica en Bolivia*, 5th ed. La Paz: Librería Editorial Juventud, 1982.

Valencia Vega, Alipio. *Geopolítica del Litoral Boliviano*. La Paz: Librería Editorial Juventud, 1974.

Vallejo V., Jorge. "Pretensión Soviética sobre la Antártica." *Memorial del Ejército* 4 (July/August 1950): 7–11.

Velilla de Arréllaga, Julia. *Paraguay, un Destino Geopolítico*. Asunción: Editorial IPEGEI, 1982.

Velilla de Arréllaga, Julia. "La Cuenca del Plata." *Diario ABC Color*, Sunday supplement (Asunción, Paraguay), June 12, 1975.

Veríssimo, Ignacio José. "Bolivia, País do Atlántico." *A Defesa Nacional* 532–533 (November–December 1958): 115–118.

Vicuña, F. *Antarctic Resources Policy*. Cambridge: Cambridge University Press, 1983.

Vignali, Heber Arguet et. al. *Antártida: Continente de los Más, para los Menos*. Montevido: Fundación de Cultura Universitaria, 1979.

Villacrés Moscoso, Jorge. *Historia de Límites del Estado Ecuatoriano*. Guayaquil: Editorial Arquidiocesiano, 1984.

Villegas, Osiris G. "El Marxismo y la Estrategia Indirecta." *La Prensa* (Buenos Aires) April 21, 1986.

Villegas, Osiris G. "Geopolítica del Atlántico Sur." Buenos Aires: unpublished monograph, May 1986.

Villegas, Osiris G. "El Conflicto Anglo-Argentino. Las Razones Aparentes y los Intereses Ocultos tras la Actitud Británica." *Geosur* 35 (1982): 5–8. (Also published in *Geopolítica* 24 (1982): 55–57.)

Villegas, Osiris G. *El Conflicto con Chile en la Región Austral*. Buenos Aires: Pleamar, 1978.

Villegas, Osiris G. *Políticas y Estrategias para el Desarrollo*. Buenos Aires: Círculo Militar, 1969.

Von Chrismar Escutti, Julio César. *Geopolítica: Leyes que se Deducen del Estudio de la Expansión de los Estados*. Memorial del Ejército de Chile, 343. Santiago: Biblioteca del Official, 1968.

Wall, P., ed. *The Southern Oceans and the Security of the Free World. New Studies in Global Strategy*. London: Stacey International, 1977.

White, Jeffrey F. *The Southern Cone and the Antarctic: Strategies for the 1990's*. Master's thesis, Center for Latin American Studies, University of Florida, Gainesville,1986.

Woolcott, R. "The Interaction Between the Antarctic Treaty System and the United Nations System." In *Antarctic Treaty System, An Assessment*, 375–411. Proceedings from the 1985 Beardmore Conference on Antarctica. Washington, D.C.: National Academy Press, 1986.

Zolezi, Daniel. "Es Indispensable una Política Austral Unica para los Conflictos del Sur." *Independencia* (Buenos Aires), July 26, 1984.

Index

About the Book

The Southern Cone of South America together with the Antarctic is a region of significant international interest, not least because of the potential and actual conflicts that abound there. There are, for example, the Falklands/Malvinas war and ongoing dispute, the tensions between Argentina and both Brazil and Chile, Bolivia's demand for an ocean outlet, and the competition for wealth in offshore and seabed minerals and for fish in bordering waters. The authors of this study, some of whose works have never before been published in English, discuss these conflicts, but point also to recent developments that suggest rising Southern Cone harmony. These developments have made the affairs of this southern region unique within international politics and important to Great Power competitors; they are indicative, as well, of a new independent force in global relationships.